Preharvest Field Sprouting in Cereals

Editor
N. F. Derera, Dip. Agr. Sci., Dip. P.B., F.A.I.A.S.
Director of Wheat Breeding (Retired)
Plant Breeding Institute
University of Sydney
Narrabri, N.S.W.
Australia

CRC Press
Taylor & Francis Group
Boca Raton London New York

CRC Press is an imprint of the
Taylor & Francis Group, an **informa** business

First published 1989 by CRC Press
Taylor & Francis Group
6000 Broken Sound Parkway NW, Suite 300
Boca Raton, FL 33487-2742

Reissued 2018 by CRC Press

Library of Congress Cataloging-in-Publication Data

Preharvest field sprouting in cereals / editor, N. F. Derera.
p. cm.
Includes bibliographies and index.
ISBN 0-8493-6848-0
1. Grain—Preharvest sprouting. I. Derera, N. F., 1919-
SB189.6.P74 1989 87-32553
633.1'049—dc19

A Library of Congress record exists under LC control number: 87032553

Publisher's Note
The publisher has gone to great lengths to ensure the quality of this reprint but points out that some imperfections in the original copies may be apparent.

Disclaimer
The publisher has made every effort to trace copyright holders and welcomes correspondence from those they have been unable to contact.

ISBN 13: 978-1-315-89687-8 (hbk)
ISBN 13: 978-1-351-07597-8 (ebk)

Visit the Taylor & Francis Web site at http://www.taylorandfrancis.com and the
CRC Press Web site at http://www.crcpress.com

PREFACE

In 1973, a group of research scientists met in Wageningen, The Netherlands, to discuss the difficulties they were experiencing in connection with research into preharvest sprouting damage in cereals. The outcome was a cooperative international program of fundamental and applied research on this subject. Since this time, our knowledge of preharvest sprouting has advanced to such a degree that CRC Press concluded that the situation warranted the commissioning of this book. It is interesting to record that two of the contributors are members of the original "team of 1973", namely, J. E. Kruger and N. F. Derera, while another, M. D. Gale, joined the cooperative soon after its founding.

In a work of this nature it is inevitable that the same subject will be discussed by more than one author, but the reader will note that rather than becoming repetitious, reports from the viewpoints of specialists such as the physiologist, the biochemist, and the geneticist add to the comprehensive nature of the publication.

This book summarizes the research done on preharvest sprouting during recent years. Naturally, whenever needed, earlier findings are also considered. Both the publishers and the authors hope that this publication will provide researchers and students with a convenient source of up-to-date information in this expanding field.

THE EDITOR

Nick F. Derera, Dip. Agr. Sc., Dip. P. B., F.A.I.A.S. is principal and founder of the "ASAS" Agricultural Science Advisory Service and former Director of Wheat Breeding, at the Plant Breeding Institute of the University of Sydney.

N. F. Derera received his training at the Royal Jozsef Nador University of Technology and Agricultural Sciences, Hungary. After years of experience as plant breeder with private seed firms, he became Director of the Hungarian Cotton Research Institute. In 1957, after the Hungarian uprising, he and his family found refuge in Australia. Here Mr. Derera worked as a factory process worker until he was appointed by the New South Wales Department of Agriculture as research agronomist. In this position he played a major role in the establishment of an economically sound cotton industry in the Namoi Valley, N.S.W. In 1961 Mr. Derera accepted a position with the University of Sydney as Wheat Breeder with the University's Wheat Research Center at Narrabri, N.S.W. He is the breeder or co-breeder of 12 wheat cultivars. He also developed improved breeding methodology and a number of screening techniques. In 1966 N. F. Derera became Officer-in-Charge of the Wheat Research Centre, and in 1973 the University appointed him to the position of Director of Wheat Breeding.

On the initiative of N. F. Derera, a series of symposia on the problems of preharvest field sprouting in cereals were commenced in 1975, giving a great impetus worldwide to research on this subject.

Mr. Derera has been Visiting Scientist in Sweden with the Swedish Seed Association and Visiting Professor at the Universities of Guelph and Manitoba. He also lectured widely in the U.S. Europe, Mexico, India, and Taiwan. He is a member of the Australian Institute of Agricultural Science (Emeritus), the Crop Science Society of America, the Australian Association of Agricultural Consultants, the Wheat Breeding Society of Australia, the New York Academy of Sciences, and the Society for Advancement of Breeding Researches in Asia and Oceania.

Among Mr. Derera's honors are: Fellow of the Australian Institute of Agricultural Science, for outstanding contribution to Australian Agriculture, the Farrer Memorial Medal for achievements in plant breeding; a plaque and citation in recognition of his research on preharvest sprouting damage; and the Rotary Award for Vocational Excellence for his contribution to the wheat and cotton industries.

N. F. Derera is the author and co-author of 83 research papers, chapters in books, and major symposia.

CONTRIBUTORS

N. F. Derera, Dip. Agr. Sc., Dip. P.B., F.A.I.S.
Agricultural Science Consultant
Agricultural Science Advisory Service
Winston Hills, N.S.W.
Australia

M. D. Gale, Ph.D.
Institute of Plant Science Research
Cambridge
England

R. W. King, Ph.D.
Senior Principal Research Scientist
Division of Plant Industry
CSIRO
Canberra, A.C.T. Australia

J. E. Kruger, Ph.D.
Research Scientist
Grain Research Laboratory
Canadian Grain Commission
Winnipeg, Manitoba
Canada

James Mac Key, Dr. Sc.
Professor Emeritus
Department of Plant Breeding
Swedish University of Agricultural Sciences
Uppsala
Sweden

Daryl Mares, Dr. Sc.
Senior Cereal Biochemist
Plant Breeding Institute
Narrabri, N.S.W.
Australia

TABLE OF CONTENTS

Chapter 1

THE EFFECTS OF PREHARVEST RAIN

N. F. Derera

TABLE OF CONTENTS

I. INTRODUCTION

In the summer rainfall and marine regions where winter cereals are grown, there is a good probability of having some rain just prior to or during harvest. This precipitation may induce the germination process in the already physiologically ripe kernel. This induced germination or sprouting triggers a sequence of physiological processes which, among others, include the evolution of a number of plant hormones and hydrolytic enzymes. A breakdown of the kernel protein and starch reserves results, which eventually degrades the kernel quality characteristics and in most cases, renders the affected kernel, unsuitable for the processing industry. The preharvest rain-damaged crop will also have a reduced yield potential and, owing to the quality defect, will attract a lower price. All major winter cereals such as wheat, barley, rye, oats, and triticale may suffer from preharvest rain damage. Preharvest sprouting damage has also been reported in rice and maize.

Slow, lasting rain followed by reduced temperature is more likely to cause preharvest sprouting than a fast shower followed by sunny periods. The grain will imbibe a soaking rain and if quick drying does not take place in lines without adequate dormancy, the germination process will be induced. The degree of damage will depend on many factors besides the temperature and the amount and intensity of rainfall, e.g., the presence or absence of genes for resistance to sprouting damage, the stage of ripeness, standing or lodging of the crop, etc.

II. EFFECT ON GROWERS

Preharvest germination in the spike causes loss in yield and also drastically lightens the test weight. Yield loss occurs partly by increased shattering and also during threshing when light-depleted kernels are winnowed out. A yield loss of 10% or more is not rare.[1] Because of preharvest sprouting, the specific gravity and the packing ratio will be reduced in the affected grain and in consequence the grain will have a significantly reduced test weight. The grower will not only suffer by having a yield loss, but will also be docked because of lower test weight and for having unwanted, sprouted kernels in the sample. In wheat acquisition systems such as the Prime Hard wheat receival centers of northern New South Wales in Australia, milling and baking qualities of the kernel are important considerations. The degree of sprouting is assessed in each load of wheat delivered. The Hagberg falling number apparatus is being used, but more recently, direct α-amylase assays have been considered. If the sample of a particular load indicates a higher α-amylase level than permitted, the load will be excluded from the Prime Hard pool and may be downgraded to the feed wheat class. Certainly this means a financial loss for the grower, as the price of the feed wheat is significantly lower than the Prime Hard wheat. This price differential may mean a loss of $15 to $40/t for the farmer.[2]

To estimate accurately the loss caused by preharvest sprouting damage to the world grain crops is very difficult; however, there is adequate information from various regions to indicate that certain crops in a particular year can be seriously affected. A 30 to 50% rain-damaged crop during a so-called bad year is not rare; . . . and, the affected kernels are not suitable for human consumption.[3] In the Prime Hard wheat production region of northern N.S.W. Australia, during a 10-year span between 1968 and 1977, aproximately 2.8 Mt of potential Prime Hard category wheat were downgraded to feed grain class. Out of this, it was estimated that at least 2 Mt could be attributed to preharvest sprouting damage. In Australia, during 11 seasons from 1972/73 till 1983/84, the average receival of preharvest sprouting damaged wheat was 7% (998,065 t) of the total wheat production per annum. In northern New South Wales and Queensland, preharvest sprouting is a recurrent problem. For the same 11-year period, on an average, New South Wales delivered 10% (470,658 t) and Queensland 16%

(159,000 t) sprouted wheat. There was a particularly rainy harvest in 1983/84. Of the total production (4.246 Mt), 20% was sprouted. Of this, New South Wales alone delivered 2.571 Mt of preharvest sprouted wheat, which represented nearly one third of the state production. In Victoria, which usually has low preharvest sprouting damage, one fifth (931,000 t) of the State production was affected by sprouting in 1983/84. At the same time, Queensland, for a change, had "only" 11% (202,000 t) of sprouted wheat.[4] These downgradings represent an annual loss of $18 million to the Australian farmers.[2]

Rye is affected by preharvest sprouting damage in a similar way to wheat. Rye is open pollinated and the glumes are more open then in wheat; hence the rain can reach the grain quite easily. Preharvest sprouting damage caused a drastic reduction in rye acreage in Sweden. Sweden used to produce 100,000 to 120,000 ha of rye per annum, but on account of a number of "bad" rainy harvest years, the resulting grain was unsuitable for milling and baking. Consequently, in 1965 the rye acreage dropped to 38,000 ha.[5] Normal production resumed when the first sprouting-tolerant cultivar, Otello, was released for commercial cultivation.

When a maturing barley crop is exposed to rain and preharvest sprouting takes place, a degradation of carbohydrate occurs. This will lead to loss of dry matter and consequently to loss of yield. The reduction of yield is the most important factor to the grower. Nevertheless, it was found that if feed barley is affected by a low level of preharvest sprouting, it will increase the digestibility of the kernel, particularly when fed to pigs. This increased digestibility usually balances out the loss of dry matter incurred due to sprouting. Some experiments indicate that care has to be taken to provide a balanced nutrition when sprouted barely is included in the feed mix.[6]

With malting barley, the brewers want a uniform, rapid germination, while the growers wish to have some dormancy to protect against preharvest sprouting in the field. These opposing requirements prevent the development of cultivars with a long-lasting dormancy, all cultivars must have only a short period of dormancy. However, even with the right dormancy, a number of kernels may have split pericarp/testa and these kernels will germinate readily if affected by rain. Pregermination as sprouting damage in most countries, particularly in the U.K., is not an important quality factor. However the wet harvest conditions in 1977 in England and western Europe forced the maltsters to give special consideration to preharvest sprouting. In most malting barley producing countries, the determination of the number of sprouted kernels in the sample is now a condition of acquisition by the maltsters. Any barley which has a large number of presprouted kernels in the sample is not suitable for malting and the grower can utilize it only for animal feed.[7] According to Jensen and Heltved, pregerminated grains should not exceed 2 to 3% in malting barley in order to avoid large losses owing to nonviable grains after storage.[8] The pregerminated grains lose their germinability during storage drying out.

Triticale, a relatively new crop, is extremely sensitive to preharvest sprouting. It may sprout in the spike without even receiving any rain. Its initial α-amylase is high. If rain damage occurs, loss in test weight and yield can be substantial. A 75% loss in yield and/or test weight after rain damage is not rare. Increased α-amylase activity on its own is not regarded as detrimental, because triticale is mainly used as animal feed, particularly for pigs.

A relatively small part of the oat production is processed for food. Preharvest rain, besides causing sprouting, may also increase shattering of the oats.

Summer cereals are not really the subject of this book, but it is worth mentioning that causes of preharvest sprouting in rice and maize differ from the winter cereals. Preharvest sprouting in rice is more frequent in the "japonica" type than in the "indica" where the grain may come in contact with water by sources other than rain. Preharvest sprouting in corn is principally attributed to vivipary.[9]

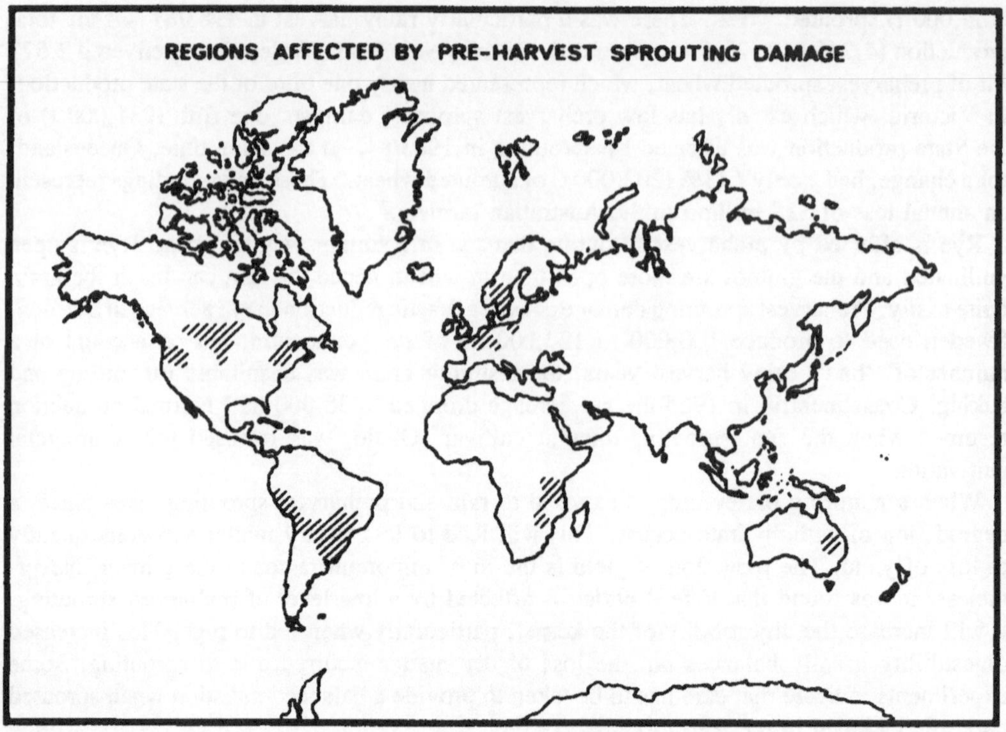

FIGURE 1. Regions affected by preharvest sprouting damage. (From Derera, N. F., *J. Aust. Inst. Agric. Sci.*, 48, 67, 1982. With permission.)

Cereals grown in northern Europe, the marine regions of western Europe, most regions of Chile, Argentina, Brazil, South Africa, Zimbabwe, Kenya, Saskatchewan and Manitoba in Canada, and eastern New Zealand are particularly liable to preharvest sprouting damage. The Pacific Northwest states of the U.S., such as Washington, Oregon, and Idaho, as well as eastern Michigan and New York, plus the Ontario province of Canada and various locations in Australia (particularly in the northern wheat belt where the most susceptible white wheats are grown) are all regions where the risk of preharvest sprouting is extremely high.[2]

It is claimed that preharvest sprouting is seldom a problem in the classical wheat-growing districts of the U.S., namely, from Texas northward to North Dakota; nevertheless, in 1977 12% of the hard red winter wheats and 19% of the durum wheats produced in North Dakota had measurable preharvest sprouting damage. North central Kansas and south central Nebraska had significant preharvest sprouting damage in the red hard winter wheat in 1979.[10]

The major regions affected by preharvest sprouting are shown in Figure 1.[11]

In Canada and the very northern U.S., a special problem occurs as a consequence of the swathing practice. While the crop is lying in windrows (swathes), and persistent rain falls, the losses owing to sprouting can be extremely high. This author experienced 80% loss on certain paddocks in Saskatchewan and Manitoba during the fall of 1978.[11]

It is well known that the wild species of wheat and their related genera have long-lasting dormancy or some other type of resistance to prevent early sprouting. Most cultivated wheat cultivars, particularly the modern dwarfs, are highly susceptible to early or preharvest sprouting. The reduction and sometimes complete disappearance of dormancy results from human intervention. Man unknowingly selected against dormancy when winter wheat was harvested late in summer and the seed derived from the same harvest was used for early autumn sowing. Only gentotypes from the early land varieties, which were able to germinate shortly after harvest, would survive. More recently, human self-interest, or actually the

FIGURE 2. The loaf on the right was baked from normal flour, while flour for the loaf on the left was derived from sprouted wheat. (Photo by C. Rowntree. From Derera, N. F., *J. Aust. Inst. Agric. Sci.*, 48, 67, 1982. With permission.)

cereal breeders' efficiency, has caused the degradation of the last traces of dormancy in spring wheats. Most modern wheat breeding programs grow two or three generations in calendar year with the help of alternation between low and high altitude sites or by growing generations on opposing hemisphere locations. Derera gave the following explanation and examples:

> CIMMYT grows a crop at Cd. Obregon on 39 m altitude between November and May and within a couple of weeks of the May harvest, the next generation crop is planted at Toluca at an altitude of 2640 m. Here it is harvested in November and the following generation is sown at Cd. Obregon and/or Toluca almost immediately. A similar system is practiced by many northern hemisphere breeding stations. They grow a generation in Europe or North America which is harvested in September/October and taken by air to New Zealand or Mexico and sown again, then harvested at the end of April or May and taken back to the original station to sow the next generation. Under this inadvertent selection pressure, only those genotypes can survive which do not possess any dormancy at harvest.[11]

III. CONSEQUENCES TO THE PROCESSING INDUSTRY

When the kernel is exposed to water and imbibition takes place, the germination process commences immediately. At first, even after slight wetting, the kernel becomes bleached and its texture turns "mealy". Considering a somewhat simplified process, the hormonal activity of gibberellic acid (GA) in the imbibed grain will increase and consequently amylases will build up fairly quickly. Owing to the increased amylase activity, the carbohydrate reserves will be hydrolyzed. The hydrolyzed carbohydrates will be translocated to and used by the growing embryo. The first symptom of preharvest sprouting damage, besides the mealy texture, is reduced test weight, then lower milling yield; also the flour has a lower Hagberg falling number (higher paste viscosity). The bread made from this flour will have high diastatic activity, giving a sticky crumb, and may produce collapsed loaves as demonstrated in Figures 2 and 3.

The proportion of sprouted grains in a sample is usually used to specify the rate of sprouting damage at the receival points. This is unsatisfactory, as the estimate of the number of sprouted grains does not correlate properly with the degree of starch degradation, which is the actual factor that the processing industry has to consider.[12] Hagberg falling number and amylograph tests are used by the cereal processing industry as the acceptable assay to detect sprouting damage or, more properly, increased α-amylase activity. There is a constant trend to search for faster analytical methods which can be used for grading sprouting damage at

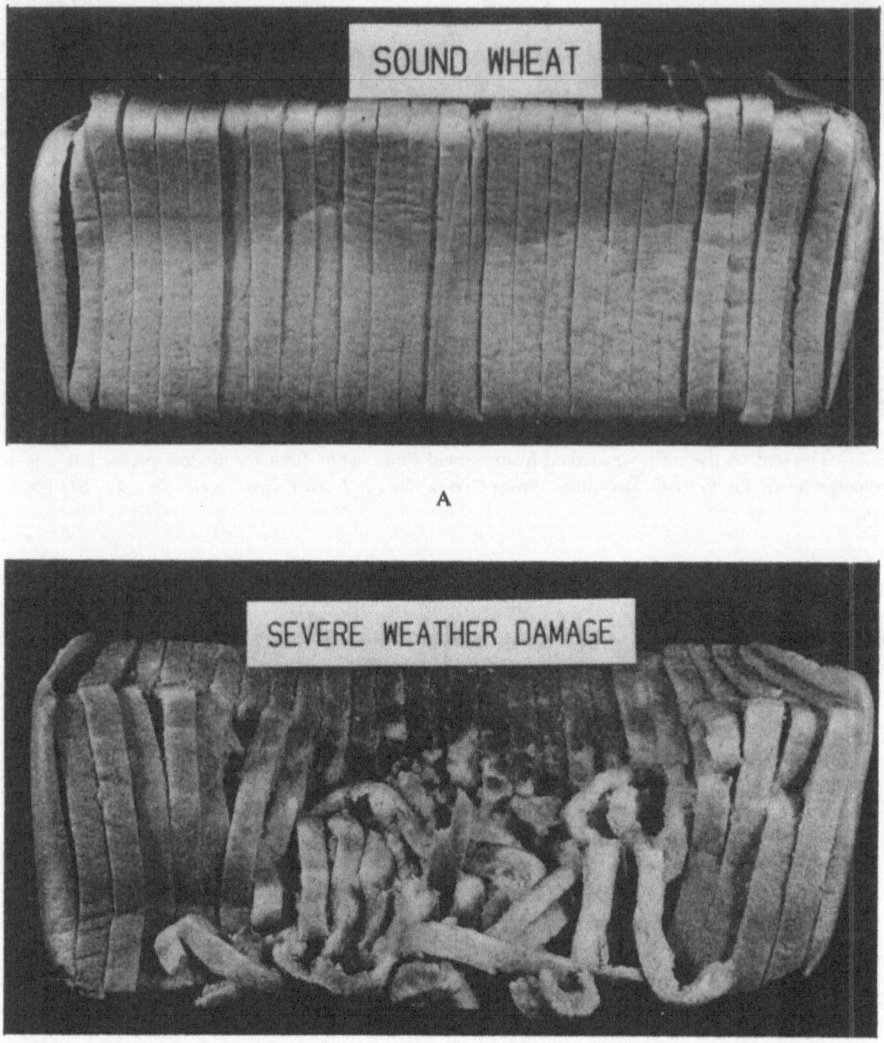

A

B

FIGURE 3. Effect of sprouting damage on sliced bread. (A) Bread baked from sound wheat flour; (B) bread baked from flour with high α-amylase content; crumb becomes sticky and the loaf cannot be sliced. (Courtesy of Mr. Stewart Haege, Bread Research Institute of Australia.)

receival points.[13] The role of the different enzymes and their assays both in research and in industry is discussed in later chapters.

Besides the amylases, the proteolytic enzymes also affect baking quality. Nevertheless, these enzymes increase slowly during germination and according to Kruger, by the time they have increased to such a level as to have an influence on bread making, the amylases will have an overriding effect.[13]

Preharvest sprouting has a deleterious effect on processing Japanese and Chinese noodles and on spaghetti manufacturing. However, in spaghetti processing, a relatively higher rate of sprouted durum wheat can be tolerated, although the color may be affected. It was found that Cantonese style noodles cannot tolerate sprouted grain in the grist. On the other hand, pan breads are more sensitive to higher levels of amylases, while chapattis and flat breads are least sensitive to sprouted grain. However, the flour produced from sound grain always results in a better product (Figure 4).[14]

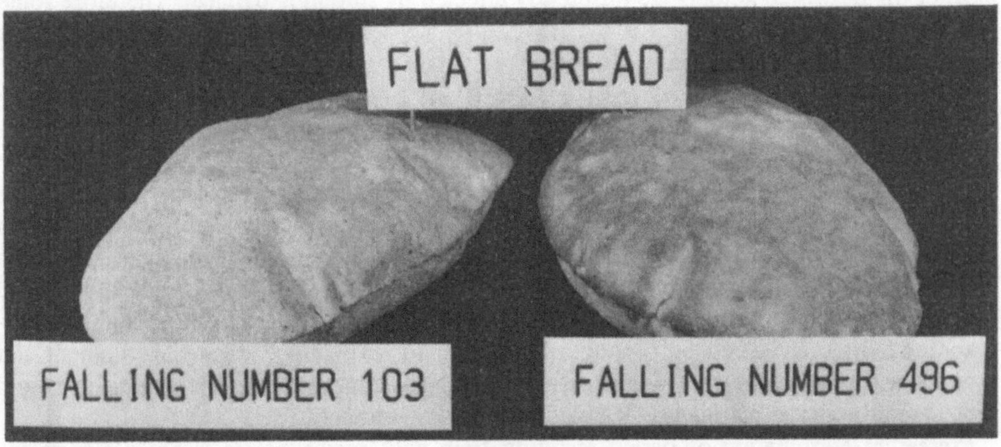

A

B

FIGURE 4. The presence of high α-amylase in the flour reduces the quality of the product. (A) Flour with low falling number produces a pale unattractive flat bread; (B) Japanese noodles made from low falling number flour will be dark and lack elasticity. (Courtesy of Mr. Stewart Haege, Bread Research Institute of Australia.)

Use of cereal-based processed food stuff is on the increase and therefore preharvest sprouting has a wide-ranging effect on the processing industry. The effects on each cereal product are different and consequently these different cereal products and their quality factors should be studied separately. It will be interesting to research how the sprouted grain can be utilized in the cereal-processing industry.

Tipples et al.[15] investigated the effect of sprouted wheat on different baking systems, namely, the Remix, Chorlywood, and continuous types and on small bakers' bread. It can be concluded that a lower starch-damaged flour can tolerate somewhat more amylase than the heavily starch-damaged one. Heavily sprouted wheat did not provide satisfactory loaves with either of the baking systems. Chamberlain reported an attempt to overcome the problems caused by sprouting damage with the development of a microwave baking system.[16] It seems that the economics of this system needs to be further studied.

In most cases, the kernel which is affected by preharvest sprouting will have a reduced density. Moreover the kernels located in different parts of the spike or kernels coming from different shoots of the plant are affected in various ways by sprouting damage. Consequently, most of the harvested kernels are differently affected by sprouting damage. Therefore some batches can be vastly improved if the low-density kernels are removed with the help of a density (gravity) sorting table.[17]

In the classical rye-producing areas of Central and Eastern Europe, rye is used primarily for bread making. Rye which has suffered preharvest sprouting damage, has, like wheat, a reduced baking quality. According to the regulation that exists in the European Economic Community, no more than 8% sprouted kernels can be tolerated before price dockage will apply. However, to be able to bake an acceptable loaf of rye bread, the rate of sprouted kernels should be below 2.5%. It is indicated that to provide reasonable baking quality the Hagberg falling number should be between 125 and 200 sec.[18] Rye bread can tolerate considerably higher α-amylase activity compared to wheat. Using the sour-dough fermentation and direct acidified processing, flours with relatively high α-amylase content can produce acceptable loaves of bread. Also, lower baking temperature and extended baking time should improve the quality of bread made from sprouted rye.[19]

Premature germination of barley resulting from split pericarp/testa is not a significant problem to the maltster. Whenever heavy rain is experienced during the preharvest period, a high degree of sprouting can be expected. Barley kernels will be reduced in density, and during steeping the affected kernels will not submerge; and there is a high chance that they will be lost in the overflow of the steeping tank. The preharvest sprouted grain will lose its germination ability and will not be converted to malt; consequently it will yield a low extract of poorly fermentable material. The preharvest sprouted kernel may develop mold during malting, which may lead to gushing during brewing.[7]

Barley samples containing 5% presprouted kernels are not acceptable for malting. It is suggested that barley in which germination has started and then been terminated by slow, natural drying, may malt quicker and provide higher grade malts than the malt produced with the normal process.[20] Nevertheless, barley so prepared could be more liable to storage defects and may not have the anticipated longevity.[21]

The maltsters require barleys with a short dormancy for immediate malting, while for later maltings, where long periods of storage are required, they prefer to have cultivars with long dormancy. Whenever the postharvest dormancy appears to be too long, the postharvest ripening can be hastened by drying the kernel at 40°C at 10 to 12% moisture. In general, warm storage will reduce dormancy. The use of GA to stimulate germination and obtain the optimum amount of malt has been used in most breweries.

To detect presprouted kernels in malting barley a number of methods were developed. Most of the tests aim to detect α-amylase activity in the kernel as the true indication of sprouting, and also attempt to obtain some quantitative assessment in the shortest possible time. Recently the Carlsberg Research Laboratory in Denmark has developed a number of ingenious systems to detect preharvest-sprouted kernels and also to monitor malting.[8,22] These methods will be discussed in Chapter 7.

IV. RESISTANCE TO PREHARVEST SPROUTING

In a broad sense, when people talk about resistance to preharvest sprouting, they associate it with dormancy.[1,23] Dormancy in the cereal kernel exists when healthy, morphologically ripe seed does not germinate when placed under optimum conditions of moisture, temperature, and light. This germination inhibition is caused by internal factors within the organ (caryopsis), and a period of "after ripening" is necessary before dormancy "breaks down".[24] Recently many researchers have found that besides dormancy there are other factors influencing, or causing resistance to, preharvest sprouting. Derera et al.[24] commented that "any exogenous factor that limits germination and factors either endogenous or exogenous to the caryopsis that limit synthesis of hydrolytic enzymes and reduce paste viscosity when dormancy

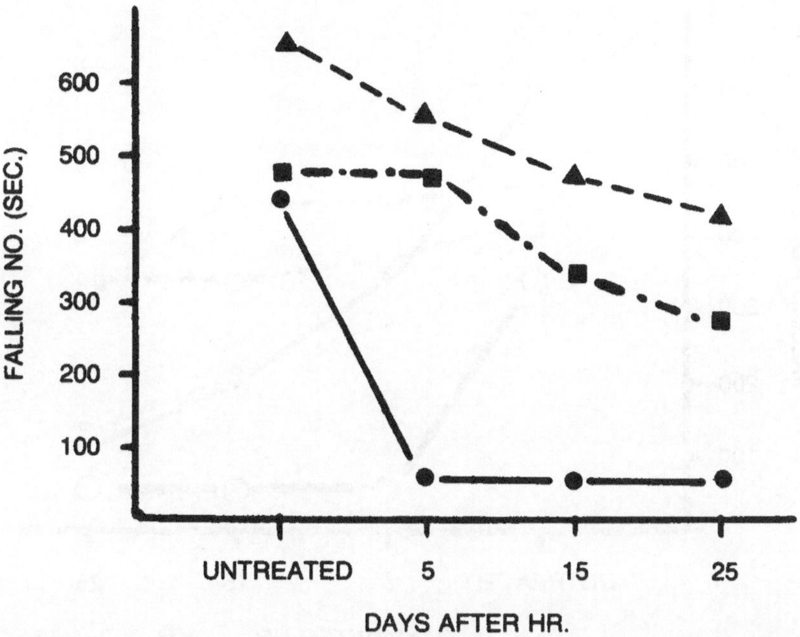

FIGURE 5. Falling number values for samples of red wheats treated with 50 mm simulated rainfall (48-hr sampling): ▲, 'RL 4137'; ■, 'Thatcher'; ●, 'Sonora 64A'. (From Derera, N. F., *J. Aust. Inst. Agric. Sci.*, 48, 67, 1982. With permission.)

does not operate, cannot be explained within the frame work of traditional definition of dormancy." These factors are numerous and the major ones which are intensively studied are water absorption/penetration to the kernel, gibberellin sensitivity of the caryopsis, amylase inhibitor proteins, and physical characteristics of the plants and spikes. For the above reasons, Derera preferred to denote the preharvest sprouting resistance as "RCP" from the Latin *robur contra pluvium*, which means resistance to rain.[11,24]

Dormancy in cereals, particularly in wheat, is usually associated with red-colored pericarp.[25-27] It must be noted that not all red wheats have dormancy (Figure 5). Some dormancy in white wheat was reported in rare recombinations by Åckerman, Voss, Isenbeck, and De Pauw and McCraig (Figure 6).[26,28-30]

The latter two authors obtained 16 white recombinations from a cross between a red spring wheat with a long, stable dormancy and a white kernel wheat without any dormancy. It must be noted that these white recombinants have a short dormancy period, which gives a much lower degree of protection against preharvest sprouting than that of the red parent. It seems that when effective protection is needed against preharvest sprouting in wheat, the solution still lies with the highly dormant wheats with red pericarps. Australia, which produces only white wheat, has a recurrent preharvest sprouting problem in the highly valued prime hard wheat-growing areas in northern New South Wales and Queensland. Derera recommended that in the critical regions Australia should change over to grow dormant red wheats instead of the susceptible white wheat cultivars.[27] The authorities and the marketing organizations did not receive this recommendation very kindly for marketing reasons. They felt that with such a change, the Australian wheat would lose its identity. Nevertheless, if Australia had changed its prime hard wheat production to highly dormant red cultivars, the growers' losses would have been minimized and several millions of dollars would have been saved.

Rate of water infiltration into the kernel may influence sprouting. Genotypic differences of water infiltration in the kernel have been reported.[31,32] The fiber content and the thickness

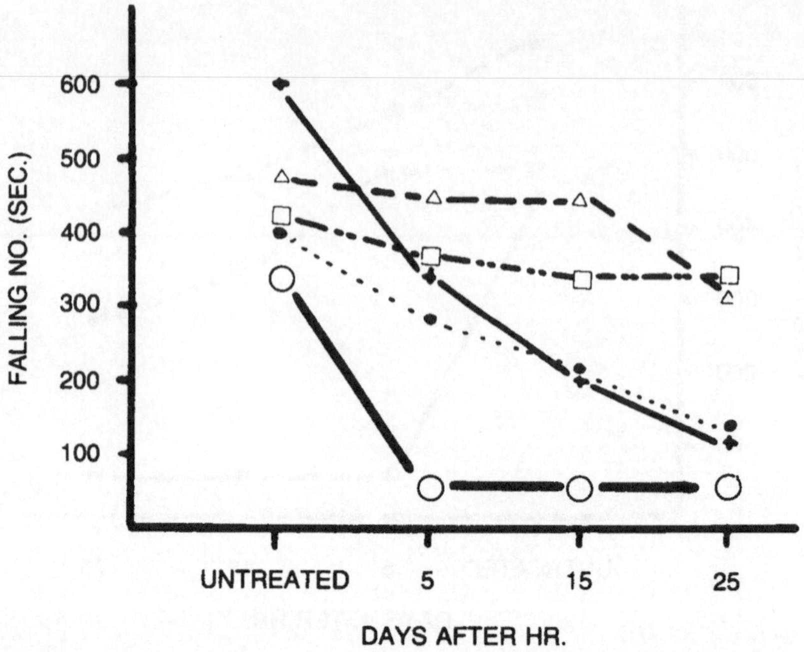

FIGURE 6. Falling number values for samples of white wheat treated with 50 mm simulated rainfall (48-hr. sampling): △, 'Kenya 321'; □, 'Ford'; ○, 'Gamut'; ●, 'Timgalen'; +, 'Tordo'. (From Derera, N. F., *J. Aust. Inst. Agric. Sci.*, 48, 67, 1982. With permission.)

of the cell wall may also influence water infiltration and the movement of α-amylase or its precursors.[33] During the drying-out process, the kernel develops minute cracks in its pericarp which enhance water and oxygen penetration and so facilitate sprouting.

Gale and Marshall described their findings that the kernels of certain dwarf varieites having the *Rht 3* gene did not respond to the exogenously applied gibberellins and their α-amylase production remained low.[34] These cultivars, which showed insensitivity to gibberellins and gave low α-amylase, were the winter wheats 'Tom Thumb' and 'Minister Dwarf.' Derefa et al.[35] reported the same gene in the Mexican spring cultivar 'Tordo' (Figure 7). At first many researchers thought that this gene might result in white wheats with potentially low α-amylase synthesis. However, it was found that the gibberellin insensitivity gene, *Gai 3*, which is present in 'Tom Thumb', is closely linked or pleiotropic with the *Rht 3* gene. Unfortunately, plants with the *Rht 3* gene are too short and are not suited for mechanized crop production systems.

α-Amylase inhibitor albumins in wheat caryopsis were reported by Petrucci et al.[36] and Buonocore et al.[37] The latter authors even suggest a breeding strategy to increase the amylase isoinhibitors in the kernel. It appears that the α-amylase inhibitors have the main role during kernel filling and the proteases break them down when the physiologically matured kernel commences germination. Intensive research is in progress on these inhibitors in wheat and barley, although the practical application of these findings is not yet clear.

It is an old farmers' allegation that certain physical characteristics of the cereal plant influence the preharvest sprouting resistance of the cereal plant, e.g., the pendant spike, the tightly closed glumes. The author of this chapter found that the pendant (droopy) spike sheds the water of a quick shower, however it does not give protection against a long-lasting, slow, so-called soaking rain. The sample applies to tightly closed glumes, as are found in speltoid wheats. The opposite is, in fact, experienced with cultivars which have so-called "gaping" glumes, particularly rye and male-sterile wheats; these absorb water quickly and

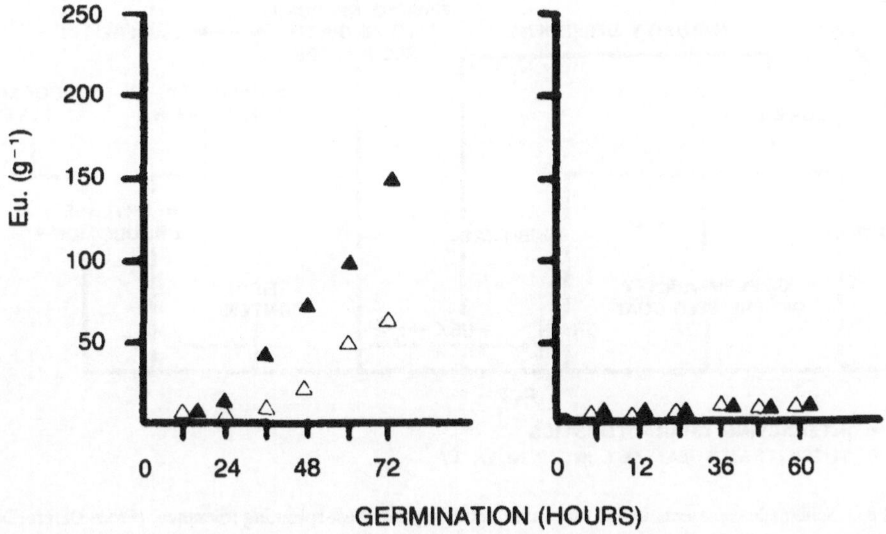

FIGURE 7. Temporal trends in α-amylase synthesis in response to GA_3 of the cultivars Timgalen (left) and Tordo (right): △, germinated in water; ▲, germinated in GA_3 ($10^{-4}M$. (From Derera, N. F., *J. Aust. Inst. Agric. Sci.*, 48, 67, 1982. With permission.)

sprout very easily if dormancy is not present. The awned cultivars imbibe water in the spike quicker than the awnless ones; also, the club-headed character increases water uptake.[38] Research conducted on the influence of physical traits of the spike has not been sufficient to draw firm conclusions.

To summarize the situation in broad terms, the sprouting resistance (RCP) factors in cereals can be divided into two major groups: ripeness-dependent and ripeness-independent components. The most important maturity-dependent factor is dormancy, which usually "breaks down" during a period of "after ripening". Dormancy may be influenced by suspected germination inhibitors, oxygen permeability of seed coat (hypoxia), and limitation of water infiltration. The last two may also act independently from ripening. In some cases the husk may contain germination inhibitors, which are also ripeness dependent. Embryo response to growth regulators is independent from ripeness. Caryopsis response to gibberellins and the level of endogenous GA has a particular importance in α-amylase synthesis. The thickness of cell walls and/or fiber content of the endosperm and the sensitivity of the starch to hydrolytic enzymes also come under the ripeness-independent group. These factors may contribute collectively and/or individually to RCP. A schematic representation of the preceding hypothesis is shown in Figure 8.

Practical experience shows that no single component of sprouting resistance will provide adequate protection in most circumstances. A combination of all resistance factors is needed to provide effective protection.[11]

V. RESEARCH

It should be noted that the occurrence of preharvest sprouting is erratic and unpredictable. It is a phenomenon which is dependent on genetical, physiological, and environmental factors and therefore is complex and changeable from year to year. There could be years of good harvests followed by cycles of rainy harvests with sizeable preharvest sprouting damage. Consequently the public and authorities are interested in the problem only during or immediately following the bad years. As soon as the weather situation changes, the general interest is dissipated. Unfortunately, this sometimes applies also to research funding. When

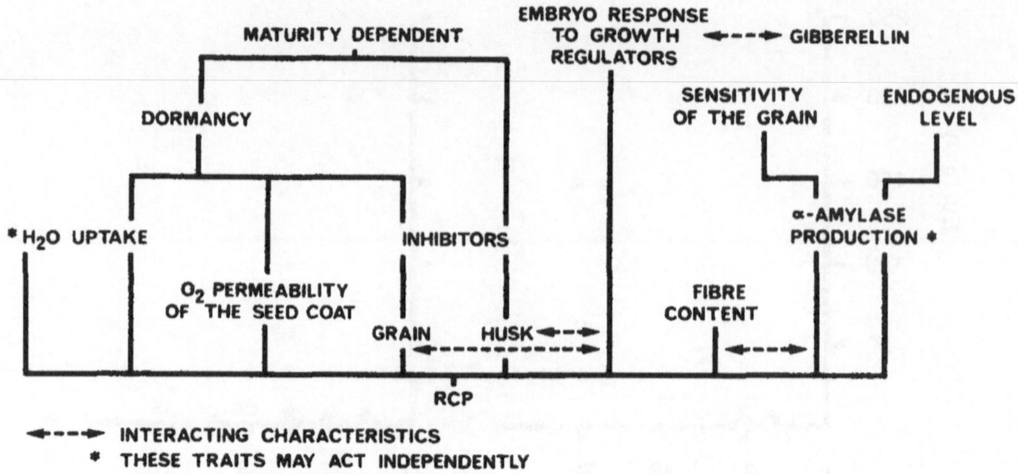

FIGURE 8. Schematic representation of factors influencing preharvest-sprouting tolerance. (From Derera, N. F., *J. Aust. Inst. Agric. Sci.*, 48, 67, 1982. With permission.)

preharvest sprouting is imminent, funds for research are provided, but when the panic is over and a better harvest cycle comes by, other seemingly "more important" projects obtain the support. It must be emphasized that this complex problem cannot be successfully researched in such a fluctuating fashion. Proper long-term planning and continuous research are needed, without panic and lull periods. While research funding of the preharvest sprouting problem is sporadic, the growers cannot expect long-term effective help from research scientists in solving or reducing this costly loss.

The preharvest sprouting and the germination process of cereal kernels have been investigated by many research workers during the last 100 years. Many scholarly papers and books have been published on the subject, and we are still far from completely understanding its fundamentals. Significant changes took place in the studies on preharvest sprouting during 1973. Prior to 1973 almost all research concentrated on dormancy. In 1973, a number of research scientists met in Wageningen, Netherlands, to initiate a cooperative venture to coordinate and mainly share findings on research on preharvest sprouting. They were B. Belderok, Holland; V. Stoy and R. Olered, Sweden; and N. F. Derera, Australia. J. E. Kruger, Canada supported the meeting but was unable to attend. This venture resulted in the organization of the First International Symposium on Pre-Harvest Sprouting Damage in Cereals which was held in Röstånga, Sweden in 1975.[39] This was followed by a second symposium in Cambridge, England during 1979, and a third symposium was held in Winnipeg, Canada in 1982.[40,41] The Fourth International Symposium on Pre-Harvest Sprouting Damage in Cereals was held in Port Macquaire, Australia.[42] These symposia created an intensified interest in the problems of preharvest sprouting and established close contacts among research workers studying different aspects of the problem. They have also generated interest in the germination process and other factors influencing preharvest sprouting. As a consequence of this increased interest, the funding of research into preharvest sprouting has improved in certain countries and is specified within the overall research allocations. As a result of this increased interest, this book on the preharvest field sprouting in cereals with particular reference to winter cereals is being published.

ACKNOWLEDGMENTS

The photographs for Figures 3 and 4 were provided by Mr. Stewart Haege, of the Bread Research Institute of Australia. Thanks are due to Mr. Doug Gow, University of Sydney,

Plant Breeding Institute, Castle Hill, for valuable assistance in preparing some of the photographic material. The author is also grateful to Ms. C. Rowntree for photographs provided. The useful suggestions and discussions on the manuscript by Dr. Norm Darvey, University of Sydney, Plant Breeding Institute, Sydney are greatly appreciated.

REFERENCES

1. **Belderok, B.**, Seed dormancy problems in cereals, *Field Crop Abstr.*, 21, 203, 1968.
2. **Derera, N. F.**, The audit of sprouting, *Cereal Res. Commun.*, 8, 15, 1980.
3. **Stoy, V.**, Progress and prospect in sprouting research, in *Proc. 3rd Int. Symp. Pre-Harvest Sprouting in Cereals*, Kruger, J. E. and LaBerge, D. E., Eds., Westview Press, Boulder, Colo., 1983, 3.
4. **McMaster, G. J.**, Pre-harvest sprouting in wheat — the Australian experience, in *Proc. 4th Int. Symp. on Pre-Harvest Sprouting in Cereals*, Mares, D. J., Ed., Westview Press, Boulder, Colo., 1987, 3.
5. **Persson, E.**, Otello — a result of amylase selection for sprouting resistance, *Cereal Res. Commun.*, 4, 101, 1976.
6. **Ringlund, Kåre**, Pre-harvest sprouting in barley, in *Proc. 4th Int. Symp. Pre-Harvest Sprouting in Cereals*, Mares, D. J., Ed., Westview Press, Boulder, Colo., 1987, 15.
7. **Brookes, P. A.**, The significance of pre-harvest sprouting of barley in malting and brewing, *Cereal Res. Commun.*, 8, 29, 1980.
8. **Jensen, S. A. and Heltved, F.**, An improved method for the determination of pregerminated grains in barley, *Carlsberg Res. Commun.*, 48, 1, 1983.
9. **Fong, F., Koehler, D. E., and Smith, J. D.**, Fluridone induction of vivipary during maize seed development, in *Proc. 3rd Int. Symp. Pre-Harvest Sprouting in Cereals*, Kruger, J. E. and LaBerge, D. E., Eds., Westview Press, Boulder, Colo., 1983, 188.
10. **Briggle, L. W.**, Pre-harvest sprouting damage in wheat in the U.S., *Cereal Res. Commun.*, 8, 245, 1980.
11. **Derera, N. F.**, The harmful harvest rain, *J. Aust. Inst. Agric. Sci.*, 48, 67, 1982.
12. **Moss, H. J., Derera, N. F., and Balaam, L. N.**, Effect of pre-harvest rain on germination in the ear and α-amylase activity of Australian wheat, *Aust. J. Agric. Res.*, 23, 769, 1972.
13. **Kruger, J. E.**, Progress in the chemistry of some quality-affecting enzymes resulting from pre-harvest sprout damage, *Cereal Res. Commun.*, 8, 39, 1980.
14. **Orth, R. A. and Moss, H. J.**, The sensitivity of various products to sprouted wheat, in *Proc. 4th Int. Symp. Pre-Harvest Sprouting in Cereals*, Mares, D. J., Ed., Westview Press, Boulder, Colo., 1987, 167.
15. **Tipples, K. H., Kilborn, R. H., and Bushuk, W.**, Effect of malt and sprouted wheat, *Cereal Sci. Today*, 11, 9, 1966.
16. **Chamberlain, N.**, Microwave energy in the baking of bread, *Food Trade Rev.*, 43, 8, 1973.
17. **Munck, L.**, The control of pre-harvest sprouting in cereals for seed, malting and milling, in *Proc. 4th Int. Symp. Pre-Harvest Sprouting in Cereals*, Mares, D. J., Ed., Westview Press, Boulder, Colo., 1987, 176.
18. **Oettler, G.**, Pre-harvest sprouting in rye and triticale, in *Proc. 4th Int. Symp. Pre-Harvest Sprouting in Cereals*, Mares, D. J., Ed., Westview Press, Boulder, Colo., 1987, 24.
19. **Weipert, D.**, Processing of grain according to alpha-amylase activity and preparing optimal blends from sound and damaged grain, in *Proc. 4th Int. Symp. Pre-Harvest Sprouting in Cereals*, Mares, D. J., Ed., Westview Press, Boulder, Colo., 1987, 197.
20. **Pollock, J. R. A.**, Studies in barley and malt. XVI. Novel malting technique involving re-steeping, *J. Inst. Brew.*, 66, 22, 1960.
21. **Sparrow, D.**, personal communication, 1986.
22. **Gibbons, G. C. and Nielsen, E. B.**, New analyses in malting and brewing *J. Inst. Brew.*, 89, 8, 1983.
23. **Harrington, J. B.**, Testing cereal varieties for dormancy, *Sci. Agric.*, 29, 538, 1949.
24. **Derera, N. F., Bhatt, G. M., and McMaster, G. J.**, On the problem of preharvest sprouting of wheat, *Euphytica*, 26, 299, 1977.
25. **Nilsson-Ehe, H.**, Zur Kenntniss der mit der Keimungsphysiologie des Weizens in Zusammenhang stehenden Faktoren, *Z. Pflanzenzuechct.*, 2, 153, 1914.
26. **Åkerman, A.**, Über die Keimungverhältnisse und Auswuchsneigung rot und weisskôniger Weizensorten, *Der Züchter*, 8, 25, 1936.
27. **Derera, N. F.**, Should red wheat be introduced into the Northern wheatbelt?, *J. Aust. Inst. Agri. Sci.*, 39, 48, 1973.
28. **Voss, J.**, Untersuchungen über Keimruhe und Auswuchsneigung von Getreidesorten (*Triticum sativum* und *Hordeum sativum* Juss.) *Landwirtsch. Jahrb.*, 89, 141, 1939.

29. **Isenbeck, K.,** Weizen: Auswuchsfestigkeit; Züchtung auf Qualität, in *Handb. Pflanzenzeucht,* Vol. 2, Kappert, H. and Rudorf, W., Eds., 1st ed., Paul Parey, Berlin, 1950, 276.
30. **De Pauw, R. M. and McCaig, T. N.,** Recombining dormancy from RL 4137 with white seed color, in *Proc. 3rd Int. Symp. Pre-Harvest Sprouting in Cereals,* Kruger, J. E. and LaBerge, D. E., Eds., Westview Press, Boulder, Colo., 1983, 251.
31. **Butcher, J. and Stenvert, N. L.,** Conditioning studies on Australian wheat. III. The role of the rate of water concentration into wheat grain, *J. Sci. Food Agric.,* 24, 1077, 1973.
32. **Stenvert, N. L. and Kingswood, K.,** An autoradiographic demonstration of the penetration of water into wheat during tempering, *Cereal Chem.,* 53, 141, 1976.
33. **Moss, H. J. and Kirby, A.,** A role for fibrous material in flour paste viscosity of wheat, *Cereal Res. Commun.,* 4, 221, 1976.
34. **Gale, M. D. and Marshall, G. A.,** The nature and genetic control of gibberellin insensitivity in dwarf wheat grain, *Heredity,* 35, 55, 1975.
35. **Derera, N. F., McMaster, G. J., and Balaam, L. N.,** Pre-harvest sprouting resistance and associated components in 12 wheat cultivars, *Cereal Res. Commun.,* 4, 173, 1976.
36. **Petrucci, T., Tomasi, M., Cantagalli, P., and Silano, V.,** Comparison of wheat albumin inhibitors of α-amylase and trypsin, *Phytochemistry,* 13, 2487, 1974.
37. **Buonocore, V., Petrucci, T., and Silano, V.,** Wheat protein inhibitors of α-amylase, *Phytochemistry,* 16, 881, 1977.
38. **King, R. W. and Richards, R. A.,** Water uptake in relation to pre-harvest sprouting damage in wheat: Ear characteristics, *Aust. J. Agric. Res.,* 35, 327, 1984.
39. **Derera, N. F. and Stoy, V., Eds.,** Proc. 1st Int. Symp. Pre-Harvest Sprouting in Cereals, *Cereal Res. Commun.,* 4 (2), 1976.
40. **Gale, M. D. and Stoy, V., Eds.,** Proc. 2nd Int. Symp. Pre-Harvest Sprouting in Cereals, *Cereal Res. Commun.,* 8 (1), 1980.
41. **Kruger, J. E. and LaBerge, D. E., Eds.,** *Proc. 3rd Int. Symp. Pre-Harvest Sprouting in Cereals,* Westview Press, Boulder, Colo., 1983.
42. **Mares, D. J. Ed.,** *Proc. 4th Int. Symp. Pre-Harvest Sprouting in Cereals,* Westview Press, Boulder, Colo., 1987.

Chapter 2

SEED DORMANCY IN WILD AND WEEDY RELATIVES OF CEREALS

James Mac Key

TABLE OF CONTENTS

I. INTRODUCTION

Already as a hunter-gatherer, man was attracted by the wild grasses which are the progenitors of modern cereals. Except for the wild rice living in swampy areas, these plants were all adapted to a climate characterized by an adverse season of the year when growth ceased. It could be due to summers being too hot and dry or winters being too cold.

In order to overcome this unsuitable period, the wild cereals developed the same strategy. They shortened their life cycle to adapt to the favorable season of the year. They could then pass the unfit period as dry, dormant seeds buried and stored in the surface soil or litter.

The annual growth habit forced them to reestablish every year in competition with other species. They had to evolve a mechanism able to start germination at the right time and they developed seeds with large endosperms, which gave immediate and plentiful access to energy at the onset of growth.

A low water content and a high concentration of energy per volume weight make such seeds not only adapted for a rapid start after a longer period of rest, but also offered man an easily storable and transportable staple food.

Even prior to the agricultural revolution, different tribes were highly dependent for their living on harvesting wild cereals. The best documented and oldest proofs are from the so-called Fertile Crescent in Asia Minor. Apart from charred seeds of typical wild appearance, primitive sickles facilitating a rapid harvest of easily shedding grain, group silos for longtime storage, and stone mills have been discovered.[1,2]

It is also in this part of the world that agriculture started. The food habit was already established and the step from gathering to agriculture was not long. Demographic reasons resulting in too intensive an exploration of the wild stands caused man to promote the spread of his favored plants by sowing and management. It was not a sudden discovery, but rather a logical solution for man with his intellectual gift. He started cultivation when it implied less labor input than to search and compete for the more and more exploited wild stands. Accordingly, we can see the same pattern often independently repeated over and over again. In the Old World. agriculture appears to have begun approximately 12,000 years ago and in the New World about 7000 years ago. Wheats and barley were chosen in western Asia, *Panicum* and *Setaria* millets in the dry and rice in the wet areas of central and eastern Asia, sorghum and *Pennisetum* millet in central East Africa, and maize in Central and South America.[3,4]

An emphasis on reestablishment essential for annual species includes also the potential for spreading. The wild cereals show their strength by invading disturbed habitats like those caused by weathering, erosion, animal grazing, fire, etc. They were thus preadapted to land prepared by man through burning, weeding, and digging. As annuals, they also had the psychological effect of giving rapid return for efforts invested.

When man through storage took over the responsibility of bridging the unfavorable season or made it favorable by irrigation, the delicate mechanism of seed dormancy lost importance and became even an annoyance. When cereal agriculture spread to more maritime climates, the need to regain some of the seed dormancy as a protection against preharvest sprouting became actual. The understanding of the dormancy mechanism in seeds of wild cereals is thus not only of academic but also of great practical concern. By this knowledge plant breeders hope to rescue great economic values by transferring the right kind of genes to modern cultivars.

II. THE STRATEGY OF WILD CEREALS

The basic principle of survival among wild cereals depends on their ability to adapt for a long period of inactivity. Their success requires the efficacy of synchronizing cycles of

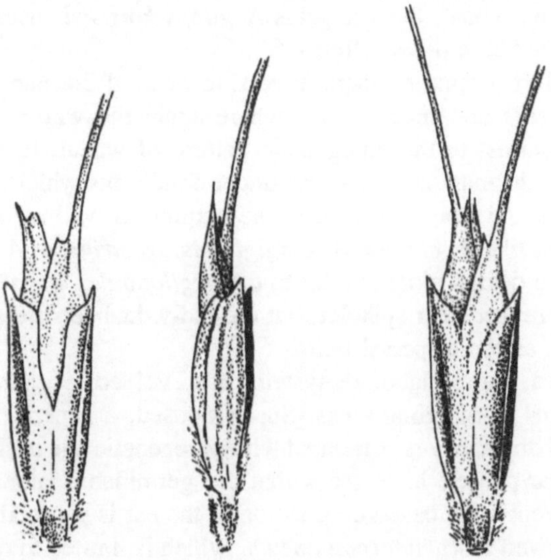

FIGURE 1. Spikelet or dispersal unit of wild einhorn wheat: (a) dorsal and (b) side view of var. *aegilopoides*, (c) ventral view of var. *thaoudar*.

growth, reproduction, and rest with changing seasons. Signals from the environment must be interpreted and used to control metabolism and development. Even the apparently inactive, dormant seed must be able to react on such outside signals.

The success cannot all be a matter of precision. Owing to fluctuations over the years in signals to follow, both the active and the passive phase must be timed with a considerable buffering. This demand for flexibility is achieved by a strategy based on opportunism, partly solved by an ontogenetic variation pattern as to the dormancy of the seeds.

Seeds from different parts of the ear/panicle or at different positions in a spikelet are given different strengths in reacting to outside signals. The pattern of how to achieve this variation among individual seeds from the same often homozygous plant may vary with species and ecogeographical specialization.

The wild diploid wheats are examples of an elaborate system. The whole spikelet, not the individual seed, is here the dispersal unit. Each spikelet has two florets the first of which sits just below and opposite the second. In the *aegilopoides* variant, the first flower is awned and fertile while the second is normally awnless and sterile (Figure 1,a,b). Each spikelet thus produces only one seed. This seed has a dormancy tuned for germination when the temperature goes down and the first rains start in the autumn. The type behaves like a winter wheat, is dependent on not being killed during the winter, and is found in the less extreme climatic zones of the Balkan Peninsula and the west and south of Turkey.

In central Anatolia and Transcaucasia, it meets the *thaoudar* variant which has two awns, two fertile florets, and thus normally two seeds per spikelet (Figure 1,c). The second of these seeds is large and has the same dormancy as the first and only seed of the *aegilopoides* type. It thus also germinates in the autumn. The basal kernel is smaller, generally only half the size, richer in protein, and darker in color. It has a lower level of endogenous gibberellin as well as a hull offering a better barrier both physically and biochemically than that of the second larger seed. Accordingly, the small seed has a deeper dormancy and is adjusted to germinate first in the following spring or even in following seasons. The *thaoudar* wheat is thus able to take advantage of an early start but also has insurance against winter killing. Its adaptation to a more trying climate is evident.[5]

A close relative to the wheat, the goatgrass *Aegilops kotschyi*, uses the same design to overcome risks for failure in a desert climate.[6]

Some oat species offer a similar adaptive version of seed dormancy. *Avena pilosa, A. ventricosa*, and *A. sterilis* are three species where again the whole spikelet functions as dispersal unit. As a contrast to the ontogenetic pattern of wheat, it is now the second or third grains that remain dormant at a time and under conditions which induce the first grain to germinate. Again a stand can be formed in the autumn as well as in the spring.

Other species of oats, like *A. clauda, A. longiglumis, A. strigosa, A. wiestii, A. barbata*, and *A. byzantina* have a design more similar to the *aegilopoides* wheat. They do, however, have more than one floret/seed per spikelet, but the individual seeds disarticulate from each other and will function as the dispersal units.

The above-mentioned, more elaborate systems are evolved to guarantee survival under great annual fluctuations in marginal areas. Superimposed, a general variation pattern operates, where degree of dormancy is correlated with ontogenetic age of the spikelet. Spikelets in the middle of the ear/panicle have seeds that are germinable later than those from the base or top. This differentiation based on position in the ear is especially well developed in wild barley, the two-rowed *Hordeum spontaneum*, which is limited to one seed per spikelet. Its ear breaks down, however, in a different way, since every three spikelets are held together forming a triplet dispersal unit.

In fact, dormancy appears never to be absolutely synchronized within genotype. Even in modern, highly homozygous cultivars, germinability revealed by preharvest sprouting follows the same positional sequence. In addition, it is well evidenced that temperature experienced by the maternal plant during seed development, and by seed following maturation, has an important, phenotypic effect on depth and length of dormancy.[8-11]

This phenotypic and largely ontogenetically controlled variation in seed dormancy is especially valuable for self-fertilizing species like wheat, oats, barley, rice, sorghum, etc. Collective homeostasis does not function smoothly enough here. Since the density of a stand will influence ear/panicle size as well as in relevant cases, number of seeds per spikelet, there will automatically be a relatively better individual safeguarding, if only few plants grow together.

In contrast to modern cultivars with high demands on uniformity, populations of wild cereals are, if not very often heterozygous, nevertheless quite heterogeneous. Heterogeneity at the individual level is thus strengthened by heterogeneity also at the population level.[12,13] Together they both offer the necessary insurance.

This does not imply that the environment-dependent selection pressure is ineffective. With a mild type of autogamy, most of the wild cereals can alternate between genotypic fixation and recombination in a very efficient way for evolutionary progress. The annual growth habit speeds up such a process by offering a rapid shift over generations.

III. THE PROCESS OF EVOLUTIONARY DIVERGENCE

The shift from gathering to cultivation proves the high adaptability of the cereal grasses. In this process, man interfered in two ways. For his own purpose, nonshattering types without deep seed dormancy became more functional.

Less consciously, man opened up another evolutionary pathway by breaking up and preparing land. Already adapted to disturbed habitats, the cereal grasses established another existence on man-cleared land. The weedy representatives of the cereal species developed. Rye and oats even started first to adapt to cultivated fields as weeds, before they were taken up as crop plants.[14,15]

The opportunistic strategy of the annual wild cereals made them just as well fitted to proceed as cultivars as weeds. In the beginning, when man was still unaware of the dynamics of life and thought plants and animals to have been created once and for all times, the selection pressure might not have been extremely different. Accordingly, we not only have weed oats and ryes but also weed wheats as well as weedy relatives within *Aegilops*, weed barleys, weed rices, weed maizes, weed sorghums, etc.[16]

It is important to understand that much of the breadth in the ecological adaptation capacity characteristic of our cereals was already present in their wild progenitors. Their particular strategy for survival had to fit rather precisely with the climatic and edaphic conditions in each niche occupied. Rice, with its adaptation to deeply flooded or swampy as well as to dry upland habitats, is perhaps the best example of adaptive breadth, in this case with water as critical factor. Wheat and particularly barley illustrate the capacity to adjust to a wide range in temperatures by abilities to climb from lowland to high altitudes. Wild oats have demonstrated their ability to spread widely into different climatic zones and daylength regimes. To some extent this flexibility is achieved by homoeostasis, a regulatory mechanism within the individual plant to buffer against changes in the environment. The real breadth is, however, achieved by ecological specialization of heterogenous populations.

It is reasonable to suggest that the wide spectrum in ecological adaptation is an accumulated result of the climatic shift in the regions concerned. When the latest glacial period ended about 20,000 years ago, a cool and humid climate was gradually exaggerated as to seasonal extremes. Climates such as the present Mediterranean type developed over much wider areas. New signals had to be followed, if seed dormancy should continue to serve as a kind of survival mechanism. Differences in altitude or other topographic variations preserved old achievements and allowed them to accumulate with new ones.

IV. SPECIALIZATION BUT PARTIAL COEVOLUTION

When man started to prepare land for the cereal grasses, their capacity for a rapid shift in evolution was apparently already available. In the beginning, the distinction between truly wild, weedy, and cultivated biotypes must have been diffuse. Gradually they became more divergent. Many taxonomists erroneously even look upon this process as a completed speciation process.

Instead, it is important to be aware of the fact that no sterility barrier exists between the three groups. A certain extent of outcrossing in a mostly autogamous reproductive system offers good opportunities for hybrid introgression and ability to take advantage of evolutionary gains outside their own group. It ought also to be considered that it is not only a question of a divergence but also of a coevolution pattern.

Both the wild progenitors and the weedy representatives are dependent on their survival by adjusting their dissemination device and seed dormancy to the ecological niche they occupy. In this respect they have tendencies in common. In their specialization, the weedy forms must, however, emphasize other morphological and physiological characteristics that make them adapted to continuously disturbed and more fertile land.

The wild relatives are, however, not uninfluenced by such trends. Where the two types coexist, the integrity of the wild form becomes more and more threatened owing to gene migration, not only from the weedy but also from the cultivated descendants. Whenever found at slopes, road banks, etc. in the vicinity of cultivated land, the genuine status of the wild type can always be questioned. Barley may serve as an illustrative example of such interwoven processes. The brittle six-rowed *Hordeum agriocrithon* has been taken both as merely a secondary hybrid derivative with weedy characters predominating and as a potential prototype.[17,18]

In features where wild and weedy relatives tend to differ, the weedy and cultivated forms instead show coevolution. Since these two groups inhabit the same type of land, occur intermixed, and can intermate, gene migration in both directions can occur more or less easily.

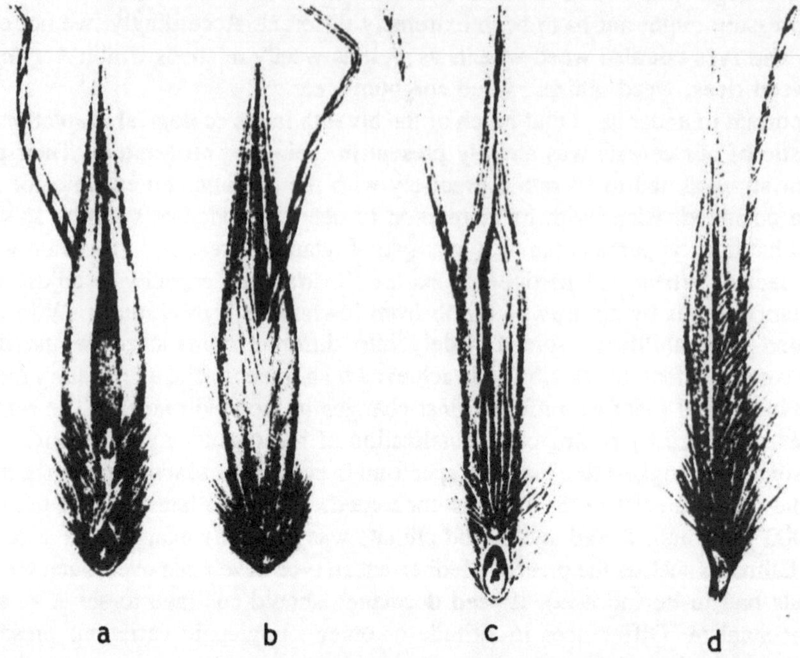

FIGURE 2. Seed and dispersal unit of *Avena fatua*, (a) ventral and (b) dorsal view; and of *A. sterilis*, (c) ventral and (d) side view. (From Stanton, T. R., *U.S.D.A. Tech. Bull.*, No. 1100, 1955, 206 pp.)

When heterogenous land races were still grown, the gene flow might have been important for improved adaptation of both types. Today a rigorous seed control blocks the direction towards the crop plant. However, the weedy relatives still have the chance to hitchhike by hybrid introgression on the crop improvements enforced by man. Since man has steadily tried to explore new land and new geographic regions, he has indirectly also promoted an evolution allowing the weedy representatives to accompany him.

V. THE CAPACITY TO INVADE NEW HABITATS

Different isolating mechanisms may intervene and control the potential gene flow. For safe survival, the weedy populations generally flower earlier than the surrounding cultivar, resulting in a kind of isolation in time. A high degree of cleistogamy, bred into the cultivated forms in order to maintain varietal purity and protect against seed-borne diseases, may increase the isolating effect.[19]

It is, however, difficult to find genetic systems directly evolved as true speciation processes. A rare case is the K_1 and K_2 genes in wheat prohibiting outcrossing with rye.[20] It is symptomatic that rye as a true weed is mostly found in wheat fields. Like other cross-pollinating species, rye has difficulties in keeping weedy and cultivated forms apart, if they grow together.

Even though there are no real barriers in cereals between weedy and cultivated relatives, the weed is not always able to spread as widely as its cultivated relative. The ability to profit on genetic hitchhiking presupposes a capacity simultaneously to adjust the seed dispersal system, including the dormancy regime.

The different capacity in this respect might be illustrated by some oat species. It shows that it is not only a matter of genetic resources but also a matter of time.

In Great Britain, *Avena fatua* is an old weed, while *A. sterilis* ssp. *ludoviciana* is a more recent introduction (Figure 2). Both were originally adapted to mature and develop seed dormancy in warm, dry summer weather with germination mainly in the autumn. *A. sterilis* is still limited to the southeastern warmest part of England and to heavy soils were winter cereals are grown.

A. fatua has a much wider climatic and, as a consequence, also edaphic adaptation. It goes far to the north and is here particularly associated with spring cereals. This noxious weed has thus accumulated a wide spectrum of seed dormancy patterns. In the north, it is adapted to cool and rainy summers, hibernation, and a germination in spring late enough to escape being wiped out at seed-bed preparation. In northwestern Europe, *A. fatua* has thus achieved regulation of its dormancy-germinability process by different signals from those originally developed in the Mediterranean type of climate.[21] Being an originally exotic weed, its efficiency has even surpassed *A. strigosa*, which as a wild plant is endemic for that part of Europe.

The seed dormancy of the weed type must not only adapt to new climatic situations. Survival on untouched as against annually ploughed or otherwise cultivated land emphasizes a different pattern of reaction. For the wild representatives, an efficient self-burial of the dispersal unit diminishes risks of loss by predation.[23] Observations indicate that up to two thirds of the seeds of wild oats and rice may be eaten by animals.[24-26]

For the weed seeds, it is more important to be able to stay alive for several years and maintain dormancy even in wet soils if buried too deep; but they must nevertheless be ready to germinate when again brought up to the surface.

A. sterilis is not able to cope with such a situation, but special *A. fatua* biotypes are.[27] They may stay alive for at least 7 years.[28] Upon the primary type of dormancy, a so-called secondary dormancy is added under conditions of excessive moisture. The right situation for germination is partly regulated by a new type of signal. At an appropriate moisture regime, the seeds are not only able to react on increased temperature but to respond also to light at onset of germination. The interrelation between water, temperature, and light appears to be very delicate and can also impose inhibition of germination.[29]

A. fatua is a remarkable example of the ability to respond to selection pressures even from such factors as different land management regimes (Figure 3). The selective advantage for types with longer dormancy could already be observed for some *fatua* populations after one summer fallow.[13] The ability to stay alive over more than 1 year becomes more important than under continuous cropping.

VI. WILD AND WEEDY FORMS AS SOURCES FOR SPROUTING RESISTANCE

All our cereals originate from regions where drought and high temperature promote seed dormancy. Under such climatic conditions, risks for preharvest sprouting hardly existed. With irrigation allowing more than one crop per year and demands on thinner hulls and testa to increase nutritional value, seed dormancy was selected against in early domestication. As agriculture spread to new climatic zones with more wet and/or cool harvest seasons, sprouting prior to harvest endangered the quality and suitability for food preparation. Introduction of combines delayed the harvest and aggravated the threat of having the grain more or less completely damaged by sprouting. Plant breeding against preharvest sprouting has thus become an important part of cereal improvement in several marginal regions.[5]

With the high degree of seed dormancy as a decisive survival criterion, the wild and weedy relatives should thus be important genetic resources. From what has been explained above, seed dormancy cannot, however, be considered an absolute factor. Its functioning is highly dependent on which type of outside signals is critical and how the environment itself influences the depth and duration of the dormancy.

This highly environmentally dependent mechanism of seed dormacy must be borne in

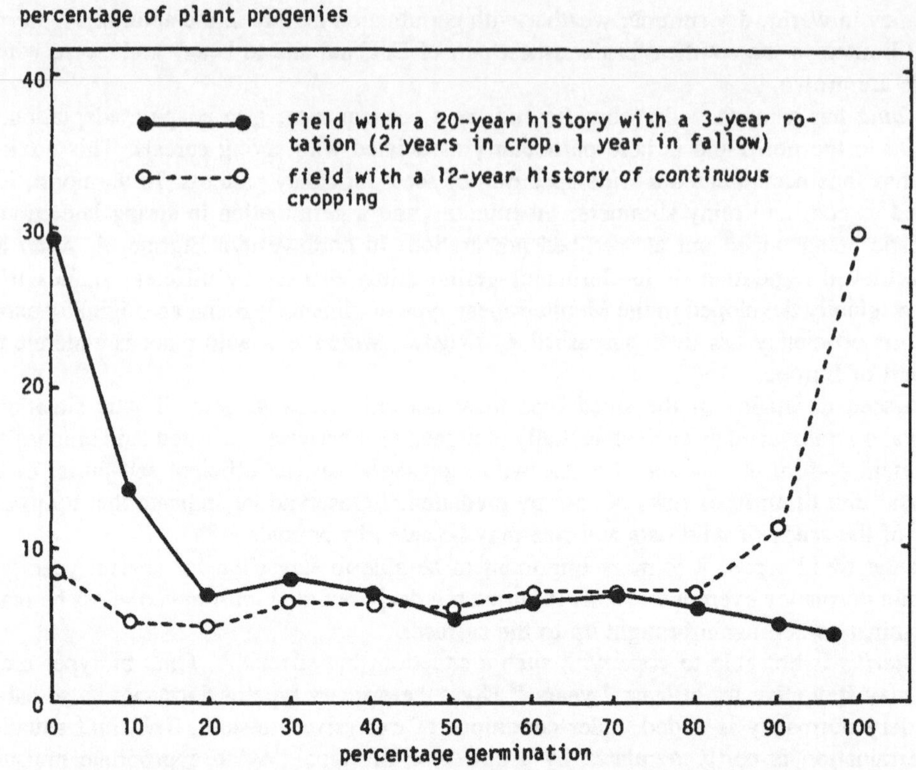

FIGURE 3. Germination rate at 3 months after maturation of seed progenies from individual *Avena fatua* plants from two fields in Saskatchewan managed in different ways (after Naylor[13]).

mind, not only when breeding for resistance to preharvest sprouting, but also at every attempt to tap the wild and weedy types for genes. They should not be considered alike nor handled each as a homogeneous group just because they all show seed dormancy. In particular, the weedy forms may represent a broad spectrum of patterns if they have been able to spread more widely than their wild ancestors. As a general rule, weedy representatives found in the same or a similar area to that for which the plant breeding program is intended should be preferred. Their pattern ought to be better tuned to the relevant situation than that of others and this includes also other features essential for overall adaptability. An appropriate exploration must consider as well the need to select in the environment concerned.

Unfortunately, the literature offers little guidance on how to deal with transfers of desirable genes for seed dormancy. In fact, the few inheritance studies almost entirely concentrated on *A. fatua* have raised more questions than they have answered. The analytical accuracy has suffered from the difficulties of relying on such an arbitrary measure as rate of germination. As indicated above in connection with a positional variation among individual seeds of the same plant, this trait shows high phenotypic plasticity. In addition, it is subject to both embryo and maternal influences as well as their interaction with the environment at three different stages of the life cycle. Tests under standardized conditions may also imply risks that certain relevant genes will become phenotypically silent and thus not be recorded.

Studies on Canadian *fatua* populations indicate a heritability around 0.5 for germination behavior.[30] In crosses between *A. sativa* and *A. fatua*, seed dormancy is found to be a recessive character.[31] It has also been found that homozygous lines within *A. fatua* may differ by at least three genes controlling rate of after-ripening.[32]

Experience from more conventional breeding programs limited to cultivars only, as well

Table 1
FACTORS INVOLVED IN REGULATING SEED DORMANCY IN CEREALS

1. Shape and surface structure of ear/panicle
2. Glume shape, tightness, and site for inhibitors
3. Seed coat and testa structure, tightness and site for inhibitors regulating water and/or gas exchange
4. Endosperm structure
5. Embryo, respiratory deficiency or blocking
6. Secondary dormancy

FIGURE 4. Different morphological versions of hexaploid wheats *(Triticum aestivum)*, varying in water-retaining capacity and thus liability to preharvest sprouting.[5]

as studies of wild and weedy forms, indicate that seed dormancy is regulated by a series of processes. Accordingly, the plant must be working with different genes or genetic blocks with distinct physiological functions triggered by such outside factors as water, temperature, oxygen, light, etc. It is quite possible that this whole regulatory system, characteristic for each biotype, is an outwork superimposed on some common respiratory deficiency or blockage associated with something like the citric acid cycle or the pentose phosphate pathway.

No matter how detailed the genetic and functional system may be, seed dormancy can almost be compared to a set of fortifications one outside the other. Table 1 sums up relevant components arranged as they successively are thought to influence the dormancy-germinability process.

For natural reasons, this whole system must be more elaborate in the wild and weedy than in the cultivated forms. Many components may not even be of interest in connection with preharvest sprouting resistance in cultivars. At least the ability to stay dormant in deep soils and react to light should have little value.

As in almost all specific plant improvement programs, breeding for resistance to preharvest sprouting will also have to face compromises. Following the listing in Table 1, ear/panicle morphology (Figure 4) has relevance to water uptake and evaporation.[33,34] Since one third of all assimilates going to the grain may come from this organ, the need for as large as possible a green surface in that region cannot be ignored.[35]

Glume or hull thickness and attachment to the caryopsis are traits where quality considerations intervene. The reason some spelting and thus non-free-threshing types of wheat are still grown in mountainous areas is undoubtedly connected with their better protection against rain showers.

A thick and tight seed coat and testa are also undesirable features characteristic of wild and weedy types. They offer both physically and biochemically important protection mechanisms against outside factors influencing germination. They lower, however, nutritional value and they impair yield and whiteness of flour. Similar industrial aspects may be laid upon endosperm structure and metabolic status partly governed by signals from the embryo itself. It should, nevertheless, be noticed that wild and weedy relatives generally have a higher protein percentage in the grain/groat compared to modern cultivars.[36,37] It is not impossible that this feature has some positive connection with dormancy.

The deep dormancy of northerly adapted *A. fatua* populations with the main onset of germination first in spring has made an insufficient winter hardiness a nonlimiting factor. This has inspired researchers to try to transfer this type of dormancy to cultivated oats.[38,39] This should allow oats to be sown in autumn in regions where winter oats cannot be grown. A fraction of the seed from such so-called dormoats may germinate in the autumn and be used as fall pasture or serve as a means of preventing soil erosion. Most of the seed should pass the winter in a dormant state and be ready to germinate early enough to take full advantage of the soil moisture. With an incomplete understanding of the complex mechanism behind dormancy/germinability, the progress of such a program has, however, been greatly hampered.

As will be evident from the following chapters, a considerable amount of knowledge has, however, accumulated. A major problem is understanding how all details should be brought together into a reliable overall system concept. This should not only help programs like the dormoats or construction of more efficient and selective herbicides against weedy cereal grasses but allow them to be better explored in breeding for resistance to preharvest sprouting in different climatic regimes.

REFERENCES

1. **Helbaek, H.**, Domestication of food plants in the old world, *Science*, 139, 365, 1959.
2. **Mac Key, J.**, Jordbrukets uppkomst [The origins of agriculture], *Skogs- Lantbr.-Akad. Tidskr.*, 119, 143, 1980.
3. **Mac Key, J.**, En tid för den förste bonden (The time for the first farmer), *Forsk. Framsteg*, 1975(1—2), 57, 1975.
4. **Cohen, M. N.**, *The Food Crisis in Prehistory*, Yale University Press, New Haven, 1977.
5. **Mac Key, J.**, Seed dormancy in nature and agriculture, in Proc. 1st Int. Sprouting Symp., Röstånga, *Cereal Res. Comm.*, 4, 83, 1976.
6. **Wurzburger, J., Leshem, Y., and Koller, D.**, The role of gibberellin and the hulls in the control of germination in *Aegilops Kotschyi* caryopses, *Can. J. Bot.*, 52, 1597, 1974.
7. **Stanton, T. R.**, Oat identification and classification, *U. S. D. A. Tech Bull.*, No. 1100, 1955, 206 pp.
8. **Sexsmith, J. J.**, Dormancy of wild oats seed produced under various temperature and moisture conditions, *Weed Sci.*, 17, 405, 1969.
9. **Sawhney, R. and Naylor, J. M.**, Dormancy studies in seed of *Avena fatua*. IX. Demonstration of genetic variability affecting the response to temperature during seed development., *Can. J. Bot.*, 57, 59, 1979.
10. **Sawhney, R. and Naylor, J. M.**, Dormancy studies in seed of *Avena fatua*. XIII. Influence of drought stress during seed development on duration of seed dormancy, *Can. J. Bot.*, 60, 1016, 1982.
11. **Reddy, L. V., Metzger, R. J., and Ching, T. M.**, Effect of temperature on seed dormancy of wheat, *Crop Sci.*, 25, 455, 1985.

12. **Sexsmith, J. J.**, Varietal differences in seed dormancy of wild oats, *Weeds*, 15, 252, 1867.
13. **Naylor, J. M.**, Studies on the genetic control of some physiological processes in seeds, *Can. J. Bot.*, 61, 3561, 1983.
14. **Vavilov, N. I.**, Studies on the origin of cultivated plants, *Tr. Prikl. Bot. Genet. Sel.*, 16(2), 139, 1926.
15. **Coffman, F. A.**, Origin and history, in *Oats and Oat Improvement*, Coffman, F. A., Ed., American Society for Agronomy, Madison, 1961, chap. 2.
16. **Harlan, J. R.**, The possible role of weed races in the evolution of cultivated plants, *Euphytica*, 14, 173, 1965.
17. **Zohary, D.**, Studies on the origin of cultivated barley, *Bull. Res. Counc. Isr.*, 9D, 21, 1960.
18. **Hammer, K.**, Bestäubungsökologische Merkmale und Phylogenie von *Hordeum* L. subgen. *Hordeum* [Characters of pollination ecology and phylogeny in *Hordeum* L. subgen. *Hordeum*], *Flora*, 175, 339, 1984.
19. **Mac Key, J.**, Significance of mating systems for chromosomes and gametes in polyploids, *Hereditas*, 66, 165, 1977.
20. **Riley, R. and Chapman, V.**, The inheritance in wheat of crossability with rye, *Genet. Res. Cambridge*, 9, 259, 1967.
21. **Thurston, J. M.**, A survey of wild oats (*A. fatua* and *A. ludovicia*) in England and Wales in 1951, *Ann. Appl. Biol.*, 41, 619, 1954.
22. **Coffman, F. A. and Mac Key, J.**, Hafer (Oats), in *Handbuch der Pflanzenzüchtung*, Vol. 2, 2nd ed., Kappert, H. and Rudorf, W., Eds., Verlag P. Parey, Berling, 1959, 427.
23. **Somody, C. N., Nalewaja, J. D., and Miller, S. D.**, Self-burial of wild oat florets, *Agron. J.*, 77, 359, 1985.
24. **Marshall, D. R. and Jain, S. K.**, Seed predation and dormancy in the population dynamics of *Avena fatua* and *A. barbata*, *Ecology*, 51, 886, 1970.
25. **Oka, H.-T.**, Mortality and adaptive mechanisms of *Oryza perennis* strains, *Evolution*, 30, 380, 1976.
26. **Cavers, P. B.**, Seed demography, *Can. J. Bot.*, 61, 3678, 1983.
27. **Whittington, W. J., Hillman, J., Gatenby, S. M., Hooper, B. E., and White, J. C.**, Light and temperature effects on the germination of wild oats, *Heredity*, 25, 641, 1970.
28. **Banting, J. D.**, Studies on the persistence of *Avena fatua*, *Can. J. Plant Sci.*, 46, 129, 1965.
29. **Hsiao, A. I.**, Mechanisms of dormancy in wild oats (*Avena fatua*), in *Proc. 4th Int. Symp. Pre-Harvest Sprouting in Cereals*, Westview Press, Boulder, 1987, 425.
30. **Jana, S. and Naylor, J. M.**, Dormancy studies in seed of *Avena fatua*. XI. Heritability for seed dormancy, *Can. J. Bot.*, 58, 91, 1980.
31. **Johnson, L. P. V.**, The inheritance of delayed germination in hybrids of *Avena fatua* and *A. sativa*, *Can. J. Res. Sect. C—D*, 13, 367, 1935.
32. **Jana, S., Acharya, S. N., and Naylor, J. M.**, Dormancy studies in seed of *Avena fatua*. X. On the inheritance of germination behaviour, *Can. J. Bot.*, 57, 1663, 1979.
33. **Varis, E. and Manneri, M.**, The effect of the morphological properties of the ear on the susceptibility of winter wheat to sprouting in the ear, *Maataloust. Aikak.*, 35, 27, 1963.
34. **King, R. W. and Richards, R. A.**, Water uptake in relation to pre-harvest sprouting damage in wheat: ear characteristics, *Aust. J. Agric. Res.*, 35, 327, 1984.
35. **Mac Key, J.**, Assimilation and yield structure — a plant breeder's conclusion, *Sver. Utsädesföeren. Tidskr.*, 94, 135, 1984.
36. **Youngs, V. L. and Peterson, D. M.**, Protein distribution in the oat (*Avena sterilis* L.) kernel, *Crop Sci.*, 13, 365, 1973.
37. **Miller, E., Levy, A. A., Avivi, L., Zamir, R., and Feldman, M.**, Evidence for maternal effect in the inheritance of grain protein in crosses between cultivated and wild tetraploid wheats, *Theor. Appl. Genet.*, 67, 521, 1984.
38. **Burrows, V. D.**, Seed dormancy — a possible key to high yield of cereals, Proc. 10th Annu. Meet. Can. Soc. Agron., Fredericton, 1964, Canadian Society of Agronomy, Ottawa, 45.
39. **Andrews, C. J. and Burrows, V. D.**, Increasing winter survival of dormoat seeds by a treatment including secondary dormancy, *Can. J. Plant Sci.*, 53, 565, 1974.

Chapter 3

PHYSIOLOGY OF SPROUTING RESISTANCE

R. W. King

TABLE OF CONTENTS

I. INTRODUCTION

At maturity the quiescent, air-dry cereal caryopsis has low metabolic activity and a water content of 9 to 12%. Then, with exposure to water, there is rapid hydration, protein synthesis may recommence rapidly, and the radicle may protrude from the seed coat within 24 to 48 hr. This protrusion of radicle and coleoptile is taken as germination. The limitations of such a description are obvious and could be avoided by using biochemical descriptors; however, there is no tightly linked sequence or network of events. Rather, many of the biochemical processes of germination occur as an essentially unrelated array involving RNA and protein synthesis and DNA synthesis, as well as mobilization of starch and other reserves. Furthermore, processes occurring in the embryo may also be spatially and temporarily separated from changes in the aleurone/endosperm tissue. To adequately accommodate such an array of response, discussion of sprouting must encompass not only visible germination, but also changes in enzymes and, in particular, of the enzyme α-amylase, since it has such dramatic effects on breadmaking and malting properties. It is necessary, also, to consider how and where enzyme levels change; are they newly synthesized, or are latent forms activated, or are inhibitors removed? In which part(s) of the grain, and when, are new mRNAs and protein synthesized? Only when such a general framework of the processes of germination is established is it then possible to seek to understand dormancy and sprouting and their environmental regulation.

II. THE DEVELOPING CEREAL CARYOPSIS

A. Grain Structure

The structure of the quiescent cereal caryopsis (termed a grain henceforth for simplicity) has been thoroughly examined by many authors.[1,2] The nonliving starchy endosperm containing both carbohydrate and protein reserves makes up the bulk of the cereal grain (Figure 1). It is surrounded by the living tissues of the aleurone layer and is separated from the embryo by the scutellum. The embryo, scutellum, and aleurone, the three living tissues, all play major roles in reserve mobilization. The two maternal tissues (integuments), the innermost (the testa — seed coat), and the outer (the pericarp), envelop the whole caryopsis. These two integuments are nonliving at maturity. In some cases, and mostly for the mature grain of barley and oats, the now-dead glumes (hull) remain adhering tightly or fused to the grain. The approximate proportion of these tissues in a well-developed wheat grain is given in Table 1.

The pericarp is colorless, the green chlorophyll-containing cells of the immature grain being part of the underlying cells of the testa. Compounds which determine coat color are also found in the testa. Below the testa is a cuticular layer (Figure 1). It has been suggested that lipid-like compounds in this layer could control uptake and loss of solutes (see Section III.B).

Until the wheat grain is almost half-mature, the maternal tissues contribute 50% of grain protein, although by maturity 90% of all protein is in the endosperm and more than 80% of this is in storage proteins.[3] Activity of enzymes including α-amylase is high in the pericarp only for about 10 days after anthesis. By maturity, in the absence of sprouting, there is little or no detectable amylase activity and these seed coat tissues are dead.

B. Grain Development

Although the isolated embryo of the half-mature, 25-day old cereal kernel may be completely germinable (see King[4,5] and references therein) maturation of the aleurone cells as determined ultrastructurally may not be complete until 65 to 70% of grain maturity.[6] In general, in the maturing kernel, germination and ability of aleurone to synthesize α-amylase

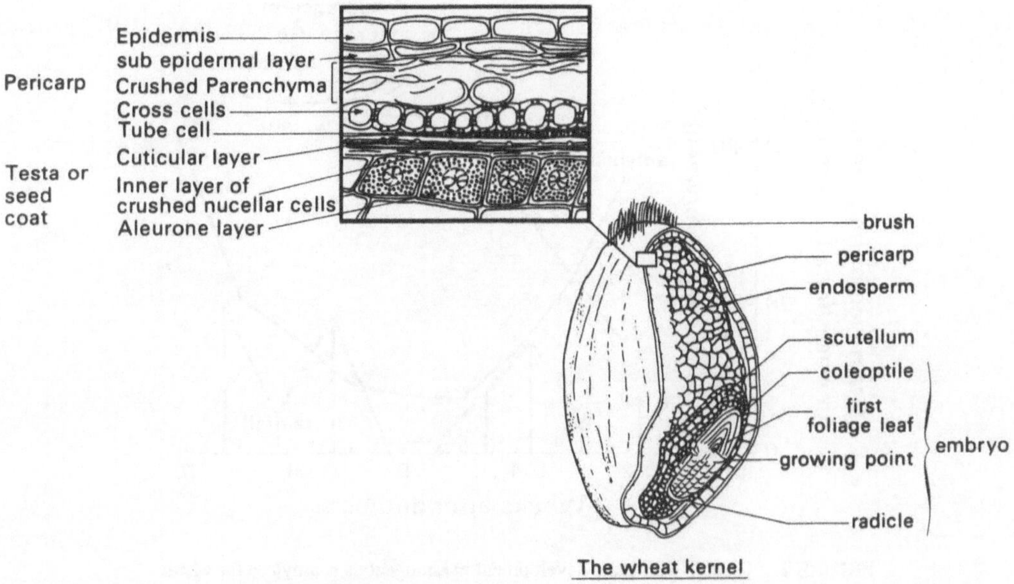

FIGURE 1. The caryopsis of wheat *(Triticum aestivum)*, showing the seed coat embryo and endosperm. Greater detail of the seed coat is shown in the inset.

Table 1
PROPORTION OF THE MAIN TISSUE TYPES IN A MATURE WHEAT GRAIN[3]

	Embryo	Starchy endosperm	Aleurone	Seed coat	Pericarp
Percentage of grain weight	2.5	84.0	6.5	2.5	4.5

are depressed until drying and cessation of grain growth. This switch is dramatic (Figure 2) and has been found not only in experiments carried out in a greenhouse but also in the field (Figure 2 and References 4 and 9). With sprouting, as in the field in 1965, there was a build-up of α-amylase induced by germination. Grain growth ceased, and its desiccation occurred at the time that α-amylase production became apparent. However, the timing of cessation of grain growth and its drying have more significance than as passive indicators of "maturity". The extent of sprouting damage is directly related to not only the timing but also to the rate of desiccation in two ways. First, if conditions are appropriate and drying is slow, then the still-moist grain can proceed to germinate, as in triticale, and a window may open briefly to allow premature synthesis of hydrolytic enzymes.[10] Little or no response is found in grain held moist, but recent studies of Gale et al.[11] highlight extensive and instantaneous response to slow grain drying with some, but not all, varieties of wheat. Second, not only the instantaneous but also the future competency to germinate and to produce hydrolytic enzymes is defined by the rate of drying of the maturing grain (normally over 3 to 10 days). As discussed later (Section IV.D), such control of future competency is reflected in both ability to respond to saturating doses of the plant growth regulator gibberellin (GA), and, possibly, to changes in the actual endogenous level of GAs and of the inhibitor abscisic acid (ABA). It may well be such changes in plant growth regulators which integrate environmental and developmental information into the response controlling sprouting.

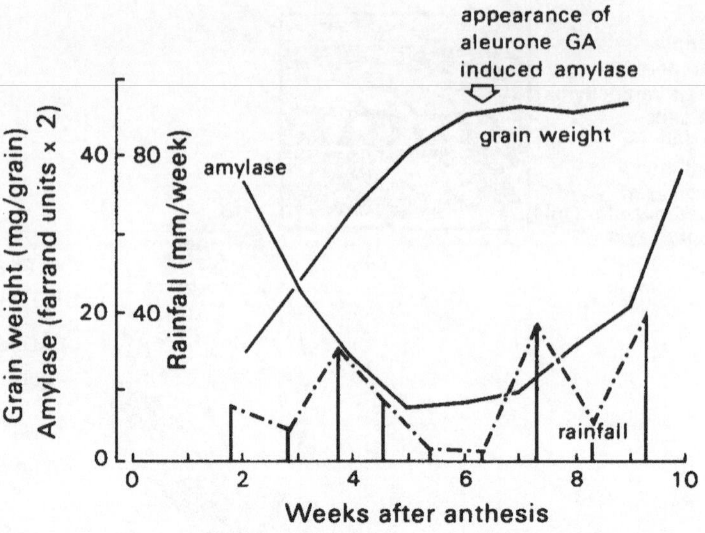

FIGURE 2. Change with grain development in endogenous α-amylase for wheat cv. Minister grown in the field at Cambridge, (U.K.), in 1965. Seasonal rainfall as indicated. Sprouting was only evident after 7 weeks. (Based upon Bingham.[8])

By contrast with the above use of quantitative measurement of grain development and maturity based on weights, there are a number of visual descriptors of maturity. There, visual estimates of ''harvest ripeness'', ''mature'' and ''ripe'', are generally poorly defined and rather imprecise, although they may be of value for varietal comparison at one site and for one season. Even so, their qualitative nature excludes important within-season differences in grain drying rates. Measurement of eosin ripeness provides a measurement which could relate to physiologically relevant processes.[12] However, although cessation of eosin uptake in the stem should reflect onset of grain drying, there has yet to be an adequate study of the relationship between ''eosin ripeness'' and true grain maturation. Further, ''eosin ripeness'' gives a qualitative, one-time measure that again gives no information on the rate of grain desiccation. Despite the ease of use for these qualitative measures of ''maturity'', if progress is to be made, they must eventually be replaced with better measures.

III. THE FIRST EVENTS OF SPROUTING

A. Rainfall and Ear Wetting

When packaged in an ear the onset of germination must be controlled by the wetting characteristics of the ear and grain. To examine ear wetting is simple with measurements of ear and grain weight changes in isolation and *in situ*. Ability to germinate can be defined from measurement of the time to attain a threshold water content for germination (see Section III.B). Alternatively, in-ear sprouting can be followed in lines held until dormancy has disappeared or, in addition, by normalizing the data following quantification of dormancy, both for isolated grain and grain in the ear. A more difficult problem is the simulation of rainfall as discussed elsewhere and by Mares in Chapter 7 of this volume.[13-14] None of the current rainfall simulators deliver particles with the kinetic energy, detachment power, and terminal velocity of natural rain. Terminal velocity is only reached at heights of about 13 m. Further, it is difficult to match simulated rain to the natural randomness of drop sizes, particularly at low intensities, and most simulators do not increase drop size and,[15] conversely, decrease drop number, as occurs with increasing intensity of natural rainfall.[16] Thus, it becomes almost impossible to simulate rainfall for all geographic locations and seasonal

Table 2
EFFECT OF AWNS ON EAR WATER UPTAKE AND SPROUTING OF WHEAT IN SIMULATED RAINFALL[13]

Ear type	Ear water uptake (%) after		In-ear sprouting (%) after
	1 hr	24 hr	30 hr
Awned 26 cultivars	25.5	46.1	39.5
Awnless 25 cultivars	20.6	43.8	21.2

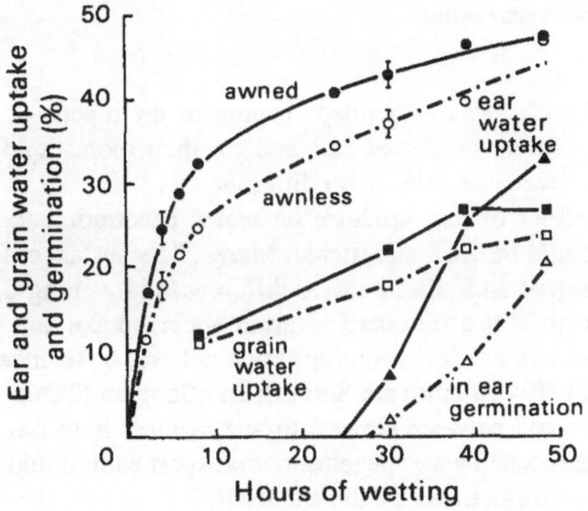

FIGURE 3. Slower wetting of the ear and of grain in the ear leads to slower sprouting in an awnless line of wheat compared with an isogenic awned line. (From King, R. W. and Richards, R. A., *Aust. J. Agric. Res.*, 35, 327, 1984. With permission.)

rainfall patterns. It is somewhat reassuring, nevertheless, that when ears of the same variety were exposed to natural rainfall for 1 hr, their wetting was as rapid as for simulated rain of the same intensity.[13]

The importance of ear morphology in limiting grain wetting in the ear and in restricting sprouting was shown recently by King and Richards.[13] Using simulated rainfall, they found that ear wetting measured at 1 hr or 24 hr and sprouting measured at 30 hr varied widely between cultivars and that some of this variation was related to the presence of awns. As shown in Table 2, awnless lines took up significantly less water and were less prone to sprouting. Similar responses were found in lines isogenic for the awn/awnless character (Figure 3). The initial differences shown in water uptake can be attributed to husk structure, since the grain was not wet until after 8 hr. Subsequently, the lag between the awnless and the awned line became even more exaggerated and when sprouting occurred, it was delayed at least 12 hr (25% slower; Figure 3).

Although it is the awned lines which are more susceptible to rain damage, the awns themselves seem to be unimportant; their removal makes no difference. Awned and deawned ears of a cultivar absorbed water at the same rate.[13] Using an isogenic awned/awnless line, the difference in initial ear wetting disappears if higher rainfall intensities are employed (25 mm/hr vs. 1 to 7 mm/hr; King[17]). Nevertheless, even with these higher rainfall intensities, grain sprouting is still reduced in awnless cultivars.[18] Thus, some other feature of husk

Table 3
EAR UPTAKE AND MOVEMENT OF WATER INTO INTACT OR ISOLATED GRAIN

Cultivar	Hr to saturation for ear wetting	Hr for isolated grain to wet to 50%	Hr for grain in-ear to reach germination water content
Songlen	3	11	24
Shortim	3	10	40
Kite	3	14	43
Sun 44E	3	16	50

Modified after Mares.[19]

architecture associated with awns — perhaps gaping of the bracts — must be responsible for both greater water capture of awned ears and for their more rapid transfer of water to the grain, even if initial water capture is no different.

Assessment of the effect of ear structure on water penetration to grain has yet to be attempted, but effects could be quite significant. Mares,[19] for instance, found that even when ear wetting was no different and, after cultivar differences in wetting of isolated grain were accounted for, there were still differences in grain water uptake and in the time taken to reach an adequate water content for sprouting. Such cultivar differences in awned lines are seen in his comparison between cultivars Shortim and Songlen (Table 3). What happens in awned vs. awnless lines and between slowest-to-wet awnless lines has yet to be examined. Clearly, however, differences in water penetration/transport paths could be major importance along with water capture differences by the ear itself.

Other ear characters warranting further examination in efforts to reduce ear wetting include ear waxiness,[20] hairiness, and the club ear gene[13] and ear-nodding angle, which significantly influences weather damage in barley,[21] although effects on sprouting were not reported. However, the effect of ear-nodding angle needs careful consideration because lax, rather than vertical ears, may be more prone to fracturing and to loss of yield. As yet, the optimal ear-nodding angle(s) for restricting ear wetting is uncertain. It has also not been documented where rain is first intercepted by an ear but, as discussed later, the pattern of formation of amylase up and down a sprouted ear points to earliest wetting of the tip of the ear.

When rain episodes are brief and, possibly, heavy, sprouting must also be influenced by rates of ear and grain drying. Ear water loss, however, appears to be simply a function of evaporation rate and is not influenced by the presence of awns in the same way as water capture.[13] Thus, varieties might all dry at the same rate and it seems reasonable then to summate each storm on the basis of the period that the grain exceeds its threshold of 45% water content for germination (see Section III.B). Germination progresses with duration, each rainfall episode being additive to previous experiences. Thus, given the restriction that, for grain viability, germination must not occur before redrying the grain, then, summation of germination becomes possible, as shown experimentally for barley and for wheat.[22,23] Rewetting may also be more rapid after a prior wetting cycle so that while not inducing visible sprouting, brief rain episodes can bring on sprouting very rapidly, apparently by accumulation of response.[23]

B. Grain Water Uptake

In the mature, air-dry cereal caryopsis, water potential may be extremely low, perhaps

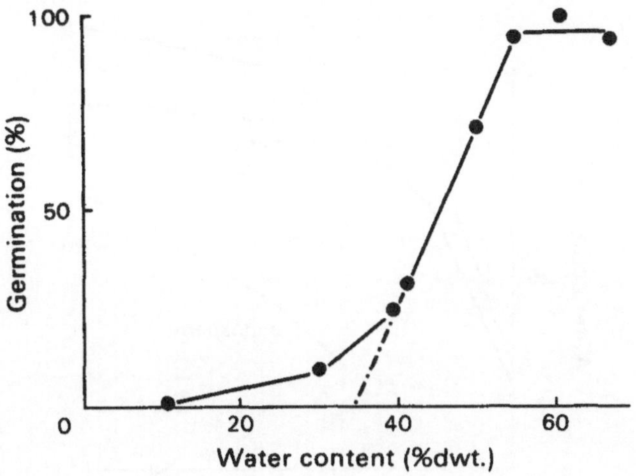

FIGURE 4. Effect of fixed levels of grain water on germination of isolated grain of wheat. (Based upon Lush, W. M., Groves, R. H., and Kaye, P. E.[22])

even as low as −100 MPa.[24] Given this extreme water potential and a kernel composed principally of nonliving starchy endosperm, then it is not surprising that water uptake is rapid (80% of maximum in 2 to 3 hr) and that, initially, physical imbibition processes dominate. In fact, living and heat-killed caryopses show the same initial rate of water uptake which confirms that, initially, only physical processes are involved.[25,26]

For many seeds, there are large early increases in volume (500%) as polymers swell in association with imbibition, but in cereal kernels the increases are small (40% in wheat) or nonexistent (oats).[27] Only later (24 hr) does total volume increase, this is associated with water absorption by the growing embryo.[28] Temperature affects initial water uptake (5% increase per 10°C), but through physical, not biochemical change. In rape, for example, the change in uptake with change in temperature can be accounted for completely by changes in viscosity.[29]

For wheat, 50% germination is only possible for water potentials greater than 0.8 to 1.0 MPa or a threshold water potential for germination corresponding to about 45% water on a fresh weight basis (Figure 4).[23,30,31] With freely available water, the critical water content should thus be reached in less than 3 hr. Effects of water potential on germination can also be assessed by using humid atmospheres. Maximal germination, however, only results when relative humidities are in excess of 97% at 20°C, so that there are limitations to this latter approach.[26]

In the cereal grain, there is no water impermeability of the seed coat akin to that found in legumes. Thus, viewed simply, when water is freely available, then varieties should differ little in water uptake by isolated grain. This is not so, however. In a sample of 49 wheat cultivars, King found differences of up to two-fold for water uptake over the first 2 hr of imbibition of isolated grain.[32] These initial rates also correlated reasonably ($r^2 = 0.37$) with germination at 30 hr. It is unlikely that chance differences in germination could have given such a correlation with a sample of this size, but it is unclear which grain characteristics could control imbibition. Using limited cultivar comparisons (2 cv.), Wellington suggested a relationship between seed coat color and grain water uptake.[28] He found that after 24 hr imbibition, the embryo of a dormant, red-grained cultivar failed to expand further or to take up more water. In these experiments dormancy was relieved by breaking the covering layers — the pericarp — and by removing the distal end of the grain. Thus, reducing the seed coat physical constraint enabled the embryo to rupture the coat. Even so, these experiments

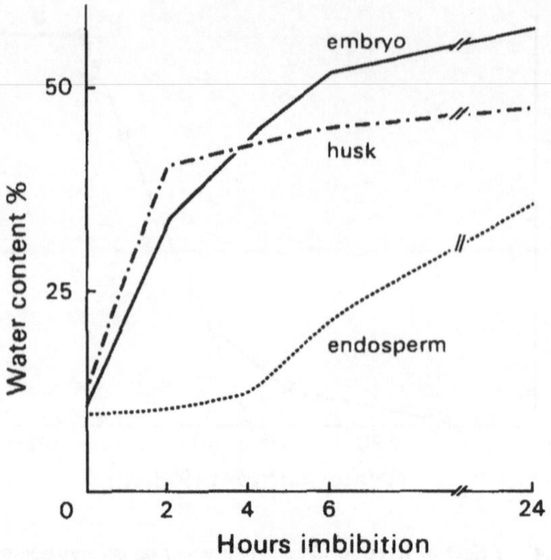

FIGURE 5. Time course of grain water uptake of individual tissues during the first hours of imbibition of barley. (Data from Reference 37.)

are not clear cut. Germination depends not only on the restrictive forces of the seed coat, but also on the ability of the embryo to generate its own turgor. Without more extensive study, either or both responses could have limited germination in Wellington's studies. By default, therefore, the questions of a mechanism of coat-controlled dormancy and water absorption is not resolved by such studies.

Grain hardness and the thickness of the lignified sub-testa suberin layer were implicated by Butcher and Stenvert and Moss in controlling grain imbibition.[33,34] However, neither grain hardness nor seed coat color influenced grain water uptake and germination in isogenic lines examined by King.[32] Also, neither grain protein content, pericarp thickness, nor the thickness of the testa or suberized layer were correlated with cultivar differences in grain water absorption or germination in wheat.[32] Grain surface-to-volume ratios could be important, and differences in this ratio between cultivars have been documented by Marshall et al.[35] However, for any one cultivar, King found differences in germination which were related to grain size (more rapid for smaller grain), but water uptake rates were no different. Embryo vs. whole grain water uptake could differ, but King did not examine embryo water uptake.[32] Surface rupturing and the formation of fissures could also differ between varieties but, despite possible water entry via these fissures, there is no evidence of varietal differences.[36]

Another possible control of grain water uptake suggested by some workers relates to the pathway of water movement. For barley, at least, it is the surface integuments (pericarp), the husk, and the embryo which first reach full hydration (Figure 5). The starchy endosperm lags considerably, and in barley,[37] at least, the lignified suberin layer of the testa (Figure 1) apparently blocks transverse movement of dyes or of aqueous GA_3.[38,39] Similarly in wheat, Hinton reported that the testa provided greatest resistance to direct transverse water uptake by grain.[40] In view of this restriction imposed by the testa on transverse water movement into the endosperm, yet given the rapid embryo hydration (Figure 5), the main path of embryo water uptake must be laterally via the pericarp to the embryo, not via the endosperm. In fact, Collins showed that the bulk of the water rapidly moves laterally to the micropyle and embryo, where the outer cuticle is also incomplete.[41] He found that sealing the micropylar end of the grain with wax blocked water uptake to a degree. For barley, support for lateral flow of water was obtained in more recent studies of Briggs and MacDonald, who supplied

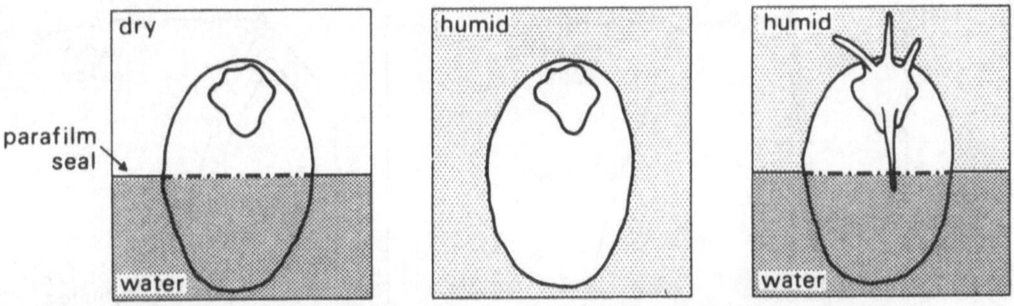

FIGURE 6. Germination after 80 hr of dehusked barley grain supplied with water to its distal half. The proximal half was maintained in either a dry or a humid atmosphere. Control grain was held in a humid atmosphere but without a supply of water. (Data from Reference 38.)

Table 4
WHEAT GERMINATION ON WET SAND

Embryo	Germination (hr to 50%)	Water uptake (%)
Up	53	60
Down	31	54

Modified after Woodbury and Wiebe.[36]

water only to the distal, nonembryo end of the grain.[38] The proximal half of the seed was held in high humidity to prevent surface evaporation of water (Figure 6), and this allowed germination. In low humidity, which should have resulted in no water flow, there was no germination. The endosperm took up little water in these experiments, but germination began when humidity was high. Hence, water was flowing predominantly up the pericarp "wick" to the embryo. The speed of this water movement in the pericarp appears to be quite slow. Woodbury and Wiebe,[36] for example, exposed the distal or the proximal (embryo half) of wheat grains to water and found a 22-hr delay in germination when there was probable lateral water movement via the pericarp (Table 4). In these latter experiments, however, lateral evaporation may not have been prevented and, therefore, slowing of water movement could have resulted.

Further experimentation is required to answer some of the questions about differences in water entry, transport, and grain wetting in the ear. As for differences in water uptake of the isolated grain, pericarp movement is unlikely to be limiting to grain on free water, and movement via the endosperm would appear to be the least favored or important route. The evidence of localized uptake of water via the micropyle and into the embryo does leave open the possibility of significant differences in water passage via the micropyle. Also, the surface of the seed coat, and its control of grain water absorption, requires further study. Possible effects of grain surface waxiness have yet to be examined in detail and with material that truly differs in surface waxiness rather than in visually assessed surface reflectivity, which can relate to wax structure rather than amount.[20,32,42]

As an aside, the germination threshold water content of 45% provides a useful measure of how varietal differences in grain water uptake influence in-ear sprouting. This initial water uptake by isolated grain should be independent of effects of dormancy on sprouting so that it would not be necessary to use after-ripened grain. Mares utilized such measurements of grain water uptake in his study of grain wetting.[19] King and Richards utilized an alternative approach of holding grain until dormancy was reduced.[13] However, there has yet to be cross-calibration of these two approaches.

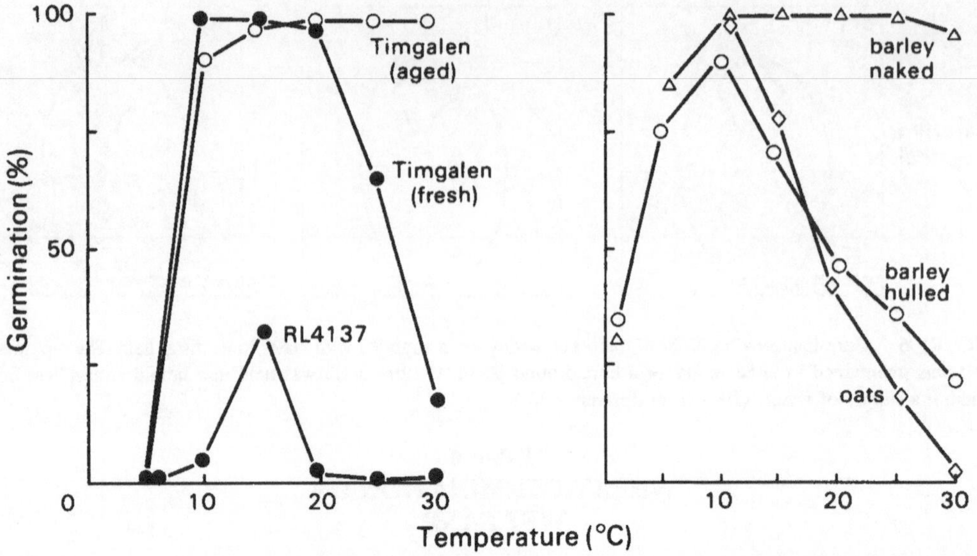

FIGURE 7. Effect of temperature on the germination of grain of wheat, barley and oats for fresh and after-ripened grain and, for barley with or without attached hulls. Data for wheat from Mares,[43] for barley from Corbineau et al.,[44] and for oats from Côme and Corbineau.[45]

IV. THE GERMINATING CEREAL GRAIN

A. Optimization of Germination

Unless germination conditions are established which are optimal for the embryo, it is easy to confound inherent differences in germinability and, perhaps, in dormancy with limitations imposed by environment. As an example, the fully ripened grain of wheat and barley apparently germinates equally well at all temperatures from 5 to 30°C (Figure 7). Is it then that the oat grain differs (Figure 7). Or, does it, too, have a flat optimum under ideal conditions? Is its reduced response at high temperature an indirect, secondary restraint? This issue cannot be resolved for oats. However, from the evidence in Figure 7 it is clear that the hull in barley and lack of seed after-ripening in wheat do impose additional limits at high temperatures and that, at least for barley, the "imposed dormancy" does not appear to restrict germination at optimal temperatures. What is not clear, however, is whether all restrictions of germination are expressed in a similar way with maximum effect at higher temperature rather than equally at all temperatures. This issue is important if seed dormancy mechanisms are to operate effectively in different environments (Section VI). Similarly, does temperature interact with further controls of germination such as are found when too little or excessive water is supplied to barley,[58] or to stimulation by KNO_3 or gibberellin? If controls of dormancy are to be examined adequately, then it needs to be known which conditions are ideal. What is optimal? Would it be preferable to work under extreme conditions? Unfortunately, there are insufficient answers to these questions.

B. Biochemistry of Events of Germination

When the cereal grain germinates, early changes are seen in proteins and RNA. Protein synthesis can be detected within 30 to 60 min of imbibition and, apparently, on preexisting mRNA and ribosomes following their reassembly as polysomes as shown by Marcus et al.[46] for wheat. Respiration also increases rapidly, in the first hours of imbibition, as found in wild oat,[47] but production of new mitochondrion occurs much later.[48] The synthesis of new mRNA and ribosomes may occur after 12 hr or more. Membrane composition may change

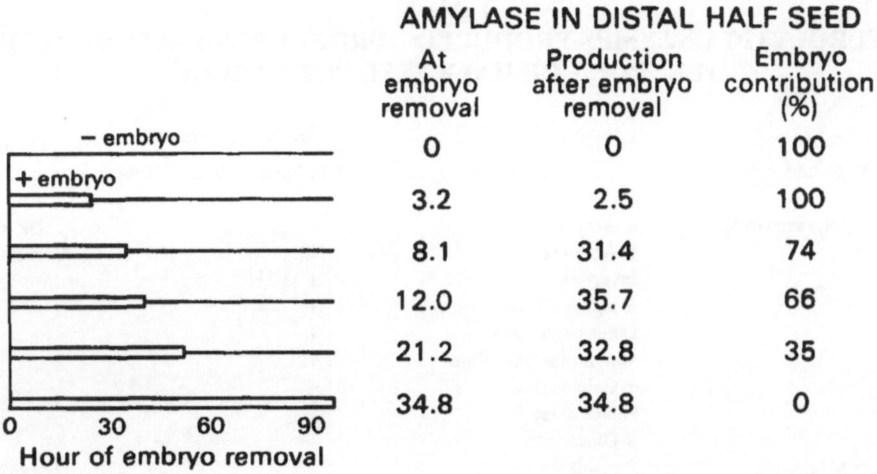

FIGURE 8. Formation of α-amylase in the non-embryo half of imbibed caryopses of barley when incubated with or without the embryo for various durations of a 96-hr total incubation period. (Adapted from Atzorn and Weiler.[50])

early on with the appearance of new endoplasmic reticulum, but its synthesis may not occur until 24 hr. In barley, at least in the aleurone, early changes in membranes reflect reorganization associated with imbibition.[49]

A critical point in germination of cereals is reached after 24 hr when damage may result if the grain is rapidly redried.[22] Apparently, it is not until cell division and early shoot and root growth commence after 24 hr that redrying disrupts the germinating kernel. Much later (1 to 2 days) there is significant mobilization of protein, carbohydrate, and nutrient reserves associated with the synthesis and secretion into the endosperm of hydrolytic enzymes including α-amylase, RNase, and proteases.[48]

While the bulk of the stored reserves in the cereal grain are found in the starchy endosperm (see Table 1), their mobilization is under the control of the embryo. In barley, for example, the embryo/scutellum is essential for 24 to 36 hr if hydrolytic enzymes are to accumulate in the nonembryo half of the grain (Figure 8). This embryo effect was noted as long ago as 1890 by Brown and Morris and is well documented for barley,[50-52] oats,[53] and wheat.[54]

C. Hormonal Regulation of Reserve Mobilization

1. Gibberellins in Mature Grain

Although the requirement for the embryo for reserve mobilization was recognized long ago, it was only in the last 25 years that it has become clear that a hormonal message,[51] a gibberellin (GA), is transported from the embryo to the living cells of the aleurone (see Figures 1 and 8). Triggering of synthesis of hydrolytic enzymes, such as α-amylase, occurs in germinating whole grain or directly in the aleurone when isolated and supplied with an exogenous source of GA to replace the embryo. Thus, in the intact germinating kernel, the embryo must be this source of GAs. In fact, recent measurements of GAs in germinating barley have demonstrated up to 20-fold increases in GAs in the nonembryo portion of the grain. In these studies, the changes in biologically active (bioactive) GAs and immunologically detected GAs were accompanied or followed a few hours later by increase in α-amylase in the grain.[50,52] Thus, it appears that GAs are an embryo factor triggering endosperm reserve mobilization.

At present, there are uncertainties about the GA putatively identified in these assays, although seedlings are known to contain GA_1 and GA_3 (see Section IV.C.3).[55] Also uncertain is the origin of the GA, whether by synthesis or conversion from a stored form, and its site

Table 5
HYDROLYTIC ENZYMES PRODUCED DURING GERMINATION OR BY ALEURONE OF HALF SEEDS OF BARLEY

Type and role	Enzyme	Hormone response: GA induces	Hormone response: ABA inhibits	*De novo*/constitutive
Starch and polysaccharide hydrolysis	α-Amylase	+	+	*De novo*
	β-Amylase	Nil	Nil	Constitutive
	Invertase	+		
	α and β-Glucosidase	+		
	Limit dextrinase	+		
	Starch phosphorylase	+		
	α-Galctosidase	+		
	Pentosanase	+		
	β-Glucanase	+	+	
Protein mobilization	Proteinases			*De novo*
	Peptidases			Constitutive
	Carboxypeptidase	Nil		Constitutive
Lipid mobilization	Neutral lipase	Nil		Constitutive
	Glyoxylate cycle and β-oxidation enzymes			*De novo*
Mineral mobilization	Phytase			Constitutive
	Peroxidase	+	+	
Other	Acid phosphatase	+	+	
	Ribonuclease	+		
	Esterase	+		

See Ashford and Gubler[1] and references therein.

of production is unknown — whether in the embryo or elsewhere in response to a further signal.[50,55] What is clear, however, is that in the aleurone there is synthesis of α-amylase in response to GAs and of a number of other enzymes important for hydrolysis of reserves as summarized in Table 5.

Four types of enzyme response are seen with GA treatment of half-seeds or on germination of barley. These are

- Activation of preexisting forms of enzymes such as those controlling metabolism of phospholipids
- GA stimulation of synthesis and secretion of an enzyme such as α-amylase
- GA-enhanced increase in an enzyme whose activity would increase even in the absence of GAs, and this includes the enzymes RNase, acid phosphatase, and β-glucanase
- GA-enhanced level of and secretion of an enzyme which would not have increased on incubation in water

For the production of α-amylase there is now thorough documentation in barley of its *de novo* synthesis involving enhanced transcription of specific genes following GA_3 treatment; these findings have been summarized recently by Jacobsen and Chandler.[56] GA_3-dependent increases in activities of a number of other enzymes also involve their *de novo* synthesis and often in an isozymic form distinct from that in the immature kernel. However, proteinases and peptidases are apparently activated on tissue hydration and not synthesized *de novo*.[57]

Despite the extensive evidence of production of hydrolytic enzymes in aleurone tissue, they are also produced elsewhere in the germinating grain. There are now a number of reports for barley,[58-62] rice, and wheat that the embryo and, especially the scutellum, accounts for the small amount of α-amylase first detected in the germinating grain.[54,63,64] These studies of the production and distribution of α-amylase are complemented by the observation that it is the endosperm adjacent to the scutellum which shows first signs of modification in barley.[65]

Not only is the embryo/scutellum producing α-amylase early in grain germination, but it may also respond to GA_3.[66] What is not clear, however, is whether the major source of this GA-induced α-amylase is from the scutellum itself or from the adjacent aleurone, which is difficult to remove when excising scutellar tissue. Palmer in his study was, in fact, able to demonstrate only limited production of α-amylase with or without GA_3 for pure cores cut from the scutellum.[66]

Overall, scutellum-derived α-amylase may account for only 5 to 10% of the total α-amylase produced in barley.[62,66] The bulk of the enzyme comes from the aleurone. Especially at early stages of germination, however, for wheat this scutellum source may increase grain content of α-amylase sufficiently to exceed a level acceptable for breadmaking. Mares,[54] for example, argues that there would be a reduction below an acceptable falling number in flour made from grain containing the amount of α-amylase that can be produced by isolated embryo/scutellar tissue. These comments require further examination, however, because even if the intact scutellum produced such an amount of α-amylase, some of this activity might normally be removed in the bran fraction on milling.

It is not known how tight the coupling is between embryo/scutellum and aleurone in their production of hydrolytic enzymes. One recent report for wheat, for instance, indicates early accumulation of amylase in the crease region of the maturing grain with no accumulation in the scutellum. In germinated grain, by contrast, early localization of amylase was, as expected, near the embryo, with change in the aleurone occurring later.[64]

2. *Gibberellins in Developing Grain*

The major GAs found in the developing cereal grain differ between species.[62] The bioactive GAs in developing wheat kernels include the less polar $GA_{9,15,20,24}$ and the somewhat more polar $GA_{19,44}$ and five C-1 β-OH GAs ($GA_{54,55,60,61,62}$). GA_{54} is present in the greatest amounts—about 10 ng per grain.[68,69] The embryo and outer pericarp contain most of the polar GAs and the less polar GAs ($GA_{15,20,24,61,62}$) may be located in the endosperm and testa.[70] By contrast, in barley the major bioactive GA is 18-OH GA_4 (maximally 200 ng per grain) and there are other 18-OH GAs, e.g., GA_{34} and 12 β-OH GA_9.[67,71] In rye there are reasonable amounts of $GA_{8,16,24}$. The major one is GA_{16} and some GA_{16}-conjugates have been detected particularly in the immature embryo.[72]

Despite the distinctions between species in their GA types a few generalizations can be made:

1. Large amounts of GAs are present in the developing grain.
2. The biological activity of these GAs is mostly low.
3. Grain GAs are distinctive from those in seedlings.[67,69]

Considering this last generalization, there have been unequivocal identifications of GA_1 and GA_3 in seedlings of wheat, barley, rye, and oats, but in the developing grain these highly active GAs are barely detectable.[68,69,73] In immature grains of wheat, for example, there is less than 0.003 ng/g fresh weight GA_1 but more than 0.2 ng/g fwt in 7-day-old seedlings.[69,74] By contrast the less active GA_{54} (30- to 100-fold less active) is barely detectable in seedlings but present at high levels in grain (170 ng/g fwt). Since recent studies with GA biosynthetic mutants in corn and pea have established that it is GA_1 which regulates seedling growth,[75] it would appear that effective GAs for control of germination and reserve hydrolysis are absent during grain growth. Thus, the differences seen between species in grain hydroxylation patterns may be unimportant; grain GAs may be essentially storage compounds.

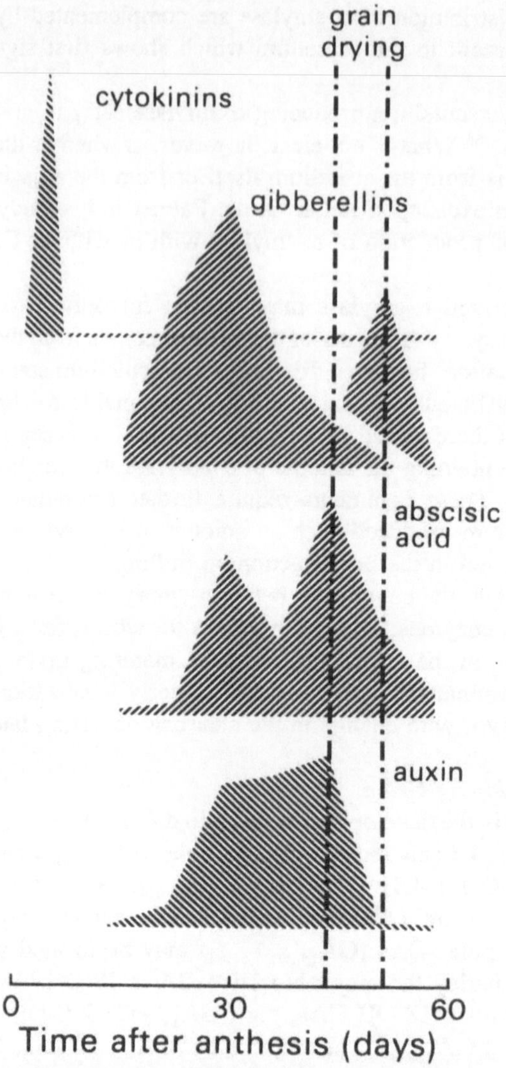

FIGURE 9. Generalized pattern of change in the content of growth regulators in the grain of wheat during its growth and maturation. (Adapted from References 4, 70, and 76.)

What is not known in relation to GAs and subsequent germination is whether with GA disappearance at grain maturity (Figure 9) there is conversion of, say, GA_{54} in wheat to some storage form for later remobilization on grain germination. For rye, the predominant GA, GA_{16}, is converted to GA_{16}-glycoside in maturing grain, but its subsequent fate is unknown.[72] GA levels in the maturing grain do, however, appear to influence germination by influencing the amount of GA that is carried through to the ripe grain. The indication of a second, late peak, of GAs in almost mature wheat grain in some of Radley's studies (see Figure 9) could indicate a carryover, although its ''new'' synthesis could have occurred with early germination.[70] No clear-cut distinctions can be made, but GA carryover does provide an explanation of Black and Naylor's evidence of release of dormancy in wild oat by GA_3 treatment of seed developing on the parent plant.[77] A similar conclusion can be applied to a study by Metzger, who found that treating the parent plant with an inhibitor of GA biosynthesis, enhanced dormancy (6 months vs. 4 months to loss of dormancy).[73] No GA could be detected in the seed from nontreated plants. Apparently, their greater dormancy reflected their lack of GAs.

Not only the amount and type, but also the location of GAs in the seed, must also be critical. Whole seed analysis may be of limited value, because the bulk of the GA might not reflect changes in tissues that contribute less than 5% of total dry matter (Table 1). The mature aleurone of corn, for instance, accumulates over 80% of radioactively labeled GA_4 or GA_{20} fed to the base of the cob near to maturity.[78] Perhaps, if sugar levels are also important through their stimulation of GA conjugation, then the pattern of GA sequestration into different tissues and pools becomes extremely complex.[79] However, it is not yet known if any of this GA "carried-over" to mature grain is significant in terms of total GAs. The only comprehensive information on this issue has been reported by Metzger with dormant/nondormant seed of wild oat.[73] He found no evidence of differences in GA content of unimbibed seed nor could he detect any significant amounts of GA conjugates. The one clear difference with regard to GA was, however, in GA biosynthesis which was at least halved in dormant compared with after-ripened seed of the same batch. This difference he was able to show by blocking GA biosynthesis and then following the decay rate of the remaining pool of GA. Such intricacies as in these experiments by Metzger serve to highlight the difficulties of making simple correlations between GA level and physiological response. Static measurements of GA content may show little of importance for responses controlled by synthesis sequestration and degradation of GAs.

3. Abscisic Acid in Mature Grain

In contrast to gibberellins, the plant growth regulator abscisic acid, ABA, inhibits germination in cereals, possibly by inhibiting wall loosening.[80] It also blocks GA-induced gene transcription and enzyme production (Table 5) while simultaneously switching on the production of specific proteins, one of which is a protein inhibitor of α-amylase.[56,81] Thus, ABA is not only capable of inhibiting GA-induced α-amylase production but, also, of "mopping-up" any existing α-amylase. The significance of such observations to dormancy and the control of sprouting is considered later. It must be emphasized, however, that, unlike the evidence pointing to GA production by the embryo, for ABA there is no evidence for a role for the embryo as a major source of ABA, nor is it known if ABA is active at the time of germination. Its action may be during germination and so reflect how much of it remains after grain maturation. Such retention of ABA may be variable and highly sensitive to environmental change (see Section IV.C.4). On the other hand, ABA present in the maturing grain may have acted previously to set limits on later germination processes.

4. Abscisic Acid in Developing Grain

Overall, the pattern of ABA change in developing cereal grain follows the pattern of dry matter accumulation (Figure 9). Thereafter, there is a substantial loss of ABA at the time of grain desiccation at maturity (see Figure 9).[82] Briefly, at this time there is an accumulation of metabolites of ABA, including phaseic acid, dihydrophaseic acid, bound forms of ABA, and an unknown highly polar metabolite.[83] General patterns of ABA change are similar for individual tissues and organs in the grain, although a change in "bound" ABA may be more significant in the embryo than the endosperm.[84]

Nonuniform distribution can occur with the concentration of ABA in the embryo being three- to five-fold greater than in the endosperm and attaining a maximum of about 10^{-7} *M*. The pericarp may contain very high concentrations of ABA during grain development, and even at maturity this may still be 10 times that of the endosperm, although now less than 25% of the total.[70] The significance of these tissue distributions in wheat is unclear, although ABA could wash from the pericarp to the embryo at the time of germination. In barley, ABA distribution in the various tissues could be of considerable importance in the malting process. For example, at maturity in dry grain, Yamada reported ABA concentrations in barley of 47, 191, and 6 ng/g dwt for husk, embryo, and endosperm, respectively.[85] Each contributed about equally (approximately 30%) to total grain ABA of approximately 18 ng/g dwt.[85] With germination, the ABA content of the embryo declined slightly, and husk ABA content dropped by three-fold over 4 days. However, on steeping for malting, grain ABA dropped by nine-fold within 20 hr and ABA appeared in the steeping liquor. Furthermore, by varying steeping conditions, it was shown that germination only occurred where ABA levels were reduced by at least five-fold.[85] Thus, the amount and distribution of ABA in mature barley could be critical in its malting.

For wheat cultivars there appear to be three- to four-fold differences in grain ABA content.[82] Walker-Simmons has reported the highest values, perhaps two-fold greater, using an immunological assay rather than a physical measurement, but it can also be expected that conditions during grain growth will contribute to such differences.[84] Water stress, for instance, elevates grain ABA content in wheat.[86] Also, for wheat grown at 25°C, ABA content was three-fold that at 15°C, and potassium deficiency may result in a three-fold greater ABA content of the grain.[70,87] The amount of ABA carried over from the immature to mature grain is also very sensitive to environment. Mature grain dried slowly (50% water loss in 79 days) maintains a high level of ABA, and its embryo will germinate poorly compared to grain dried at normal field rates (50% water loss in 8 days) and for which ABA may drop eight-fold.[82] ABA levels remain high with very rapid drying (50% water loss in less than 1 day) and, again, low germination results in association with high final ABA content.[88]

There are indications in the results summarized above of a rough inverse relationship between ABA content and germinability. More precise data on material grown at one location have also shown that ABA levels are higher (up to 50%) in less germinable cultivars of wheat and barley.[84,89,90] In contrast, in wild oat, where germination inhibitors are considered important (see Section V.C), no differences could be found in ABA content either of mature grain of a dormant vs. a nondormant line or with loss of dormancy in time.[91] Although ABA need not control all forms of dormancy, this and other negative findings[92] require some consideration. There are also, however, methodological shortcomings. For example, in response where dormancy is known to be induced by either or both embryo and integuments, then tissue localization of ABA and differences with development need examination. In wheat, for instance, the developing embryo may lose its germinability briefly (5 to 10 days) when ABA is highest in the immature grain (Figure 10).[82] If this early change in germinability in immature grain is essentially a switch for later dormancy, then whether or not ABA differs in the mature grain may not be critical. All effects of ABA on dormancy may have been completed prior to maturity. Recent studies of two ABA-induced proteins, an amylase inhibitor and wheat germ agglutinin,[81,93] support such a postulated early action of ABA in immature grain. The level of these proteins correlates with ABA build-up during grain development. Furthermore, applied ABA triggers their production. Thus, ABA is acting prior to maturity to inhibit germination, possibly to block the production of germination-associated enzymes such as α-amylase and, also, to induce embryo specific proteins and, in particular one protein, an amylase inhibitor, which is antagonistic to germination processes. An extreme but contrary example of such ABA control during grain growth is seen in the viviparous germination of some corn mutants and in fluridone-treated lines in which there is little or no detectable ABA.[82,94,95] Similarly, it would appear to be the reduction in ABA content in the developing seed of some mutants of *Arabidopsis* which leads to the mature seed being less dormant.[96] Mutants of this type, such as are found in *Arabidopsis*, have yet to be identified in wheat or barley.

Overall, a number of roles for ABA can be considered in its regulation of germination in cereals and these include

- Maintenance of seed (embryo) development rather than its germination

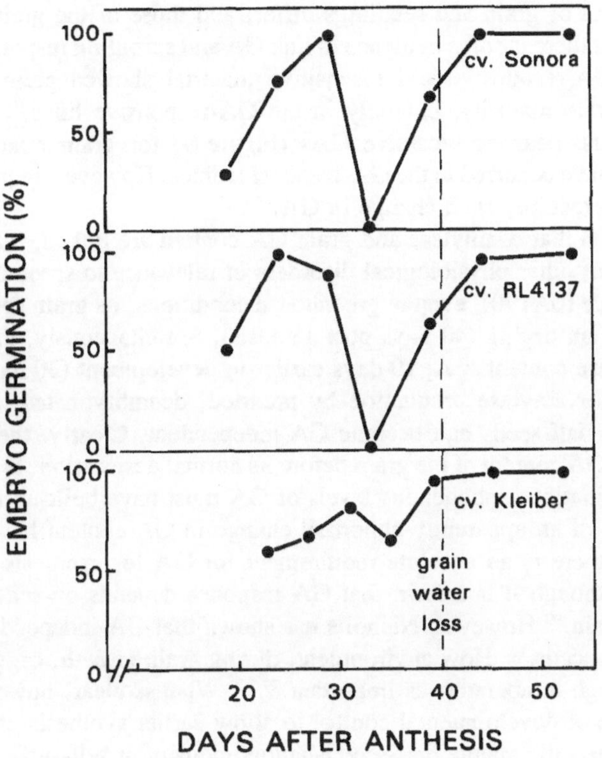

FIGURE 10. Effect of developmental stage of the embryo of wheat on its germination when isolated and placed on moistened filter paper (From King, R.W., in *The Physiology and Biochemistry of Seed Development*, Khan, A. A., Ed., Elsevier, Amsterdam, 1982, 157. With permission.)

- Initiation of dormancy in the immature seed so giving a dormant state at maturity
- Maintenance of dormancy in the mature seed
- Blocking germination, halting the production of hydrolytic enzymes

D. General Implications for Grain Sprouting

Considering the patterns of accumulation of the four major types of growth regulators in the developing grain of wheat (Figure 9), the early transient peak of cytokinin may be important in early cell division and in determining final grain size. No physiological significance for sprouting has been attached to the accumulation of free auxin during the period of maximum grain growth, although Radley noted that more auxin was found in grain of a nondormant than a dormant cultivar of wheat.[97] For abscisic acid and gibberellin, their maximum content occurs late in grain development and so may be relevant to studies on grain sprouting. In the broadest sense, this proposal reflects the known response of the mature grain to gibberellin, either from the embryo or applied exogenously to trigger production of hydrolytic enzymes and to promote germination. Likewise, for ABA it is its inhibition of germination and of GA action and its triggering of the synthesis of particular proteins which implicates ABA in the control of grain sprouting. ABA would negatively reinforce these changes in ability to respond to GAs. The high level of ABA in the immature grain would prevent precocious germination and so establish dormancy by its immediate action. Alternatively, by its carryover into the dry grain, ABA could maintain/induce dormancy at maturity.

Although the GAs of grain and seedlings differ, and those of the grain may be physiologically less active, there is some evidence to link GA and sprouting responses. For instance, a comparison of GA-responsive and insensitive material showed premature build-up of amylase at or prior to maturity, but only in the GA-responsive lines.[74] Possibly the late build-up of a second peak of bioactive GAs (Figure 9) for grain maturing at a lowish temperature might have occurred in the GA-responsive lines. However, in preliminary studies Lenton could not detect any such change in GA_1.[98]

On the assumption that α-amylase and grain GA content are linked, it has been possible to probe a number of other physiological disorders of relevance to sprouting. For example, in one line of triticale (6A190), even in greenhouse conditions, its grain sprouts prematurely in the ear just at grain drying (40 days post anthesis). Simultaneously, there is a dramatic increase in α-amylase content.[10] At 10 days earlier in development (30 days post anthesis), GA was required for amylase production by predried, deembryonated, half-seed. By 40 days, however, the half-seeds had become GA independent. Clearly, these findings point to a late increase in GA content of the grain before its normal desiccation. With uncoordinated or precocious germination, endogenous levels of GA must have built-up in the kernel.

Another instance of an apparently abnormal change in GA content has been reported in barley. Normally, there is an absolute requirement for GA for synthesis of hydrolases by barley aleurone, although it is known that GA response depends on season, cultivar, and effects of age of grain.[99] However, Nicholls has shown that GA-independent production of α-amylase can also occur.[147] How environment, during grain growth, triggers this switch is not yet clear although temperature is important.[29,100] What is clear, however, is that there has been no change of developmental control to allow earlier synthesis of a latent amylase or a precursor. Rather, the mature half-seed is autonomous of or self-sufficient for GA. This latter explanation of grain self-sufficiency fits with the report that its GA content may be elevated by ten-fold.[100] Similar indications of GA self-sufficiency are evident in wild oat. Dormant lines of this species require GA for synthesis of α-amylase by half-seeds, and GA treatment of the maturing grain on the parent plant can overcome this block.[77] Nondormant lines, however, can range from GA dependence to independence or self-sufficiency for amylase production.[101] Similar GA-priming of α-amylase production is seen in corn. The isolated aleurone of corn normally produces as much α-amylase in the absence as in the presence of GA_3. However, it can be made GA dependent if endogenous GA levels are reduced as in dwarf, low-GA mutants.[102]

Loss of GA control in lines overproducing amylase (as above) is probably the most definitive indication of excessive levels of GA. Dormancy, by contrast, would reflect an insufficiency of GA and low GA content. However, many differences in amylase production may also result from a change in the capacity of the tissue to respond to GA. One such instance is seen in the appearance at maturity of sensitivity of the aleurone to saturating doses of GA (Figure 2).[4,103] At this time, very large differences in synthetic capacity can be established, depending on the rate of grain drying over the final days before maturity.[9] In this case, by using a range of humidities at a constant temperature to impose differences of four- to five-fold in grain drying rate, then subsequent potential for amylase synthesis could be varied over a six-fold range.[9,104] Actual synthetic potential (rate) was not altered, only the incubation lag which was lengthened from 24 hr to 48 hr before first detection of amylase synthesis in response to applied GA_3.[9] The α-amylase synthesis was GA_3 dependent but, since the exogenous GA dose was saturating, then the GA content cannot explain the response to drying. Considering the suggestion of Gale and co-workers of GA involvement in the late "sprouting" sensitive window of α-amylase build-up in wheat (see above and Gale et al.[11,74]) it is clearly possible that a change in sensitivity to GA rather than a change in GA content could explain their findings. Furthermore, many other environmental and edaphic factors could be expected to alter capacity to produce α-amylase without directly altering GA or its initial perception. Differences in nitrogen nutrition, as a simple example, have been shown to alter not only grain protein but also to increase amylase synthesis in wheat and diastatic activity in barley (see references in King[5]) and, more recently, protein content has been shown to correlate positively with sprouting.[105]

Although a change in GA content or in perception/sensitivity to GA must be important, a lowering of ABA content of grain could equally well explain many of the increases in grain amylase discussed above. With the appearance of GA-responsiveness on drying of wheat grains there is, for instance, a reduction in grain ABA content for grain dried artificially when harvested immaturely or for fully developed grain allowed to dry naturally.[4,83] The ABA content of slowly dried grain also drops in an inverse manner to its GA sensitivity in wheat.[82] However, the build-up in amylase prior to maturation in *Triticale* 6A190, as noted above, appears to be unrelated to grain ABA content.[10]

In summary, GA and ABA act antagonistically on germination and their roles in sprouting include

- Applied GA promotes germination; ABA is inhibitory.
- Endogenous and applied GAs trigger embryo-controlled secretion of hydrolytic enzymes; applied ABA is inhibitory.
- Lines with differences in potential for α-amylase production may also differ in GA dependence.
- Loss of GA-dependency may result in kernels with enhanced GA content or with premature germination, while grain with low ABA content may germinate precociously.
- Applied ABA blocks actions of GA and triggers its own specific proteins including an amylase inhibitor.
- Endogenous levels of ABA correlate inversely with embryo germinability and, also, ABA content at maturity may parallel dormancy, possibly via direct effects of grain drying rate on ABA content.
- Some aspects of dormancy may be explained as a GA inadequacy or an over-production. The converse is suggested for ABA.

V. DORMANCY

A. General Concepts

If there is little or no dormancy in the mature cereal grain, it can be disastrous when conditions favor grain sprouting. Dormancy, however, is a complex control system and is difficult to manipulate in breeding programs. Also, some inconvenience occurs when highly dormant lines are to be resown or, for barley, used for malting. Another major difficulty is that the level of dormancy, once established, may show considerable instability due to seasonal conditions so that even slight differences in maturation timing could completely erase what was considered stable dormancy. In this regard, high temperatures reduce dormancy but, if grain becomes wet after maturity, then a few days of cool temperatures can break all dormancy.[43] How environment influences dormancy, its development, maintenance, and breakdown is discussed later. Here, control mechanisms in the seed are examined, those controls which render nongerminable a seed exposed to optimal conditions of moisture, temperature, and light for its germination.

The major dormancy mechanisms include:

- Embryo immaturity
- Seed coat/husk imposed dormancy: chemical controls; physical constraint; oxygen restriction
- Embryo-related processes: chemical inhibitor; lack of promotor; respiratory control; synthesis regulated; membrane properties

Table 6
EFFECT OF SEED COAT PIERCING ON GERMINATION OF WHEAT

	Germination (%)
Control:dormant grain	6
Piercing above embryo	85
Piercing then washed 1 hr	45
Piercing, washed 1 hr, then pierced again	75

Data from Belderok.[115]

As discussed elsewhere, in wheat, embryo immaturity may be unusual (e.g., in 1 of 100 wheat cultivars studied by Gordon)[106,107] although Meredith and Pomerantz claim that immaturity will be widespread at higher latitudes.[108] Light responsiveness via photoperiod and phytochrome is not important in the grain of cereals.

B. Seed Coat/Husk Imposed Dormancy

1. Chemical Controls

Considering chemical regulation of dormancy, this is a multifaceted response. The clearest evidence is for hulled grain which may fail to germinate but are completely germinable on removal of the hulls. Furthermore, substances inhibitory to germination can be found leaking into solution during imbibition or in aqueous hull extracts of oat, *Avena sativa*,[109] *A. fatua*,[53] and barley, *Hordeum vulgare*.[110] Various inhibitory chemicals have been identified in these hull extracts including salicylic acid, vanillin, free acids, and triterpenoid glycosides. There was no evidence in the study of Häggquist and co-workers for involvement of phenolic substances.[109] For wheat, the seed coat itself may contain water-soluble inhibitory compounds, and a comprehensive assay of their content in the almost mature testa indicated a contribution to total inhibitors of about 50% by catechin- and tannin-like compounds, 30% by alkaloids, and 20% uncharacterized.[111] The more dormant red-grained cultivars contained almost twice as much catechin-tannin-like compounds as found in the white cultivars.[111,112] As yet uncharacterized inhibitors of germination have also been detected in aqueous extracts of husks of some wheat and barley cultivars, but there is some question about the reproducibility of this response across seasons.[110,113,114]

Despite this evidence for control of germination by hull and coat inhibitors there is also some evidence of water soluble promotors. For wheat, for example, Belderok found that piercing the seed coat allowed germination (Table 6) but this promotion was only partially related to release of physical restraint.[155] A chemical promotion was also evident in response to wounding. For instance, if the grain was washed for 1 hr following piercing, then its germination was reduced, apparently by removal of a water soluble, piercing-induced promotor (Table 6). Repiercing the grain reinstated this response.

2. Physical Constraint

Physical constraint as a mechanism of seed coat imposed dormancy has been inadequately studied. The early studies of Wellington comparing a dormant red- and a nondormant white-grained cultivar highlighted inability to continue embryo expansion after 24 hr water uptake for dormant grain.[28] However, this study did not address questions of differences in ability of the embryo to break through the seed coat rather than in resistance offered by the seed coat itself.

3. Oxygen Restriction

Inhibition of germination by restriction of oxygen availability is a further role of hulls,

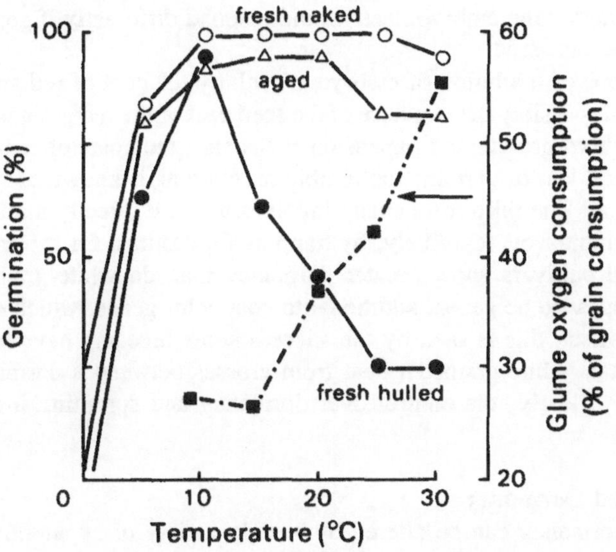

FIGURE 11. Effect of temperature on grain germination and on oxygen consumption by hulls of barley. (Extensively modified from Côme and Corbineau.[45])

at least in barley. Côme and co-workers,[116] for example, found that the hull tissues consumed oxygen and this consumption paralleled the activity of polyphenoloxidase. For grain with intact hulls, the effect on oxygen availability was greatest (most oxygen consumed) at temperatures inducing most dormancy (Figures 7, 11). At 30°C the hulls consumed up to 50% of the total oxygen consumed by the grain (Figure 11).[117] Neither aged grain nor fresh, naked grain showed this temperature-induced dormancy of imbibed grain. Moreover, after the "aging" process, the hulls of aged grain consumed much less oxygen (Côme and Corbineau;[45] 436 vs. 154 $\mu\ell$ O_2 fixed g/hr for dormant and nondormant hulls, respectively). If the hulls are inducing dormancy at higher temperatures by depriving the grain of oxygen, then it would seem that it is a deficiency of oxygen over the earliest hours of germination which is critical. Enhanced oxygen consumption by hulls was evident only during the earliest times (first 10 hr) of imbibition, while for dehulled grain, peak oxygen consumption was delayed until 20 to 30 hr of germination.[118] Considering this evidence in its totality, in barley and possibly oats also, the hull could play an important roll in dormancy by either or both trapping oxygen and release of an inhibitor.[44]

In wheat with no hull as an oxygen trap, but still with high temperature dormancy (Figure 7), no simple conclusions can be drawn about effects of seed coat. Physical constraint and or inhibitors could prevent germination,[28,111] but oxygen consumption appears to be unrelated to dormancy.[119] Oxygen consumption during the first hours of imbibition (10 hr) is also no different for dormant and nondormant cultivars of wheat,[111] although germination can be promoted at elevated oxygen levels.[120]

As an aside, it has long been recognized that white-grained wheat varieties are more susceptible to sprouting than are the red-grained ones.[121] The control of coat color by at least three distinct genes is now well known, but their effects on sprouting have yet to be comprehensively compared in isogenic backgrounds. Nevertheless, it is clear, particularly from studies of Freed and co-workers, that the presence of genes for red seed coat color correlates with post harvest dormancy.[122] They used seed of F_1 crosses of red × white and white × red grain, which gave them a standard heterozygous embryo genotype, but with the maternal seed coat tissue being homozygous and either red and "dormant" or white and "germinable". Whether the embryo itself would respond differently if genotypically red or white has yet to be answered.

Three mechanisms of inhibition of embryo growth by the coat of red grained wheats can be considered. First, possibly the structure of the seed coat of the red grain is more restrictive, physically, for embryo growth and expansion.[28] Second, the coats of red-grained cultivars could prevent natural loss of germination inhibitors. Nothing is known of such a mechanism in red-grained wheat. The third mechanism involves the coat directly inhibiting the embryo by providing an inhibitor or, less likely, by trapping O_2 destined for the embryo. However, not all red-grained cultivars show greater dormancy than do white-grained wheats. Furthermore, there appear to be genes, additional to coat color genes, which control dormancy and sprouting in wheat; this is seen by the success some breeders have had in developing a reasonably dormant white-grained wheat from crosses between a dormant red × white-grained cultivar.[123] Clearly, the control over dormancy and sprouting involves more than seed coat color.

C. Embryo-Linked Dormancy

Since embryo dormancy can be altered by a wide variety of apparently unrelated environmental and chemical treatments, it is difficult to identify any particularly dominant control(s). Furthermore, the question of what induces dormancy becomes confused when it is realized that the mechanism of loss of dormancy may or may not be related to the mechanism of induction of dormancy. Action of some chemical or environmental signal at one time may have set in train a sequence of developmental changes all of which maintain dormancy. Conversely, dormancy could be induced and maintained by the continued presence of a factor whose gradual disappearance leads to appearance of germinability. A complication is that the embryo, at least of wheat, shows simply heritable differences in sensitivity to inhibitory chemicals; these genotypic differences could be quite subtle biochemically.[124]

1. Chemical Inhibitors and/or Lack of Promotors

Most evidence of chemical regulators of embryo dormancy relates to applied hormones, such as ABA and GA, and to natural products, such as coumarins and phenolics.[106] When applied to cereal grain, these compounds will inhibit germination, except for GA, which is promotory. For ABA, recent measurements (see references in Reference 82) have shown a build-up in the ABA content of the grain of wheat at or before the time it enters dormancy. Furthermore, sprouting susceptible cultivars of wheat and barley have higher levels of grain ABA just before or just at the first appearance of dormancy.[89,90] More measurements are needed of embryo contents alone; this is now possible given the sensitivity and rapidity of immunoassays as applied by Walker-Simmons and Sesing.[125] An overview of present evidence of the role for ABA in dormancy, particularly in relation to precocious germination and control of aleurone function, was presented earlier. The relationships remain correlative. However, compelling evidence for a role for ABA in initiating dormancy has come in mutation experiments with *Arabidopsis* which showed a simple inheritance of nondormancy; these seeds had low contents of ABA.[96] Conversely, for mutants low in GAs and with reduced germination a further mutation for low ABA levels removed dormancy.[126]

Were the action of ABA to induce dormancy, as a once only response just prior to maturation of the cereal grain, then it could be unimportant that levels of ABA do,[84,89,90] or do not,[4,91] differ in the mature dormant/nondormant grain. Even so, both early action and a continued role for ABA can be envisaged. For gibberellins, by contrast, it may be that the GAs in the maturing grain are not only ineffective but also at low levels in the mature grain. Thus, the only time their promotive action would become important would be on their synthesis/release on imbibition of mature grain. The GA/ABA balance would thus become of relevance only at germination. The only GA measurements for dormant vs.

nondormant cereal grain support this scenario. Metzger found no differences in the endogenous GAs of dormant versus after-ripened seed of wild oat.[73] However, on imbibition, GA turnover was more rapid in the after-ripened, nondormant seed so their retention of GA at similar levels in the two seed types points to a real enhancement of GA synthesis at germination of the nondormant seed. Overall, some control of dormancy in cereals might be explained by ABA and GA and their interactions, but more evidence is required. Furthermore, effects of husk and hull inhibitors must be integrated into the responses, particularly for the grain subject to sprouting in the ear in field conditions. The weakness of these suggestions is, however, that they do not indicate how the duration of dormancy is regulated. For instance, were ABA important but retained at a constant concentration, then release from dormancy would have to reflect gradual realignment (i.e., change in sensitivity to ABA) of a "dormant" condition of metabolism to eventually give a nondormant grain. Such changes are not unreasonable but are totally speculative.

An alternative explanation of how the duration of dormancy is measured, introduces a chemical which is progressively lost (decays) with time or, the converse, one whose synthesis builds up with time. Berrie and co-workers argued that changes in volatile fatty acids with time provided one such decay controller of dormancy.[91] They showed that short chain fatty acids inhibited germination of wild oats and were present at inhibitory levels in the kernel. They were slowly volatilized over a period of months when stored as the pure compound[127] and, for the C_7, C_8, and C_9 fatty acids, their endogenous level declined 10- to 20-fold over the period of 6 months that reflected loss of kernel dormancy.[91] However, a more recent study of Metzger and Sebesta[127] with a different field population of *A. fatua* failed to confirm these findings of Berrie et al.[91] Metzger and Sebesta were able to detect fatty acids but not changes in levels.[127] The critical result, however, was that while applied nonanoic acid inhibited germination they found, as did Berrie et al.,[91] an abrupt switch from noninhibition to inhibition of germination, but this change was apparently due to toxicity rather to induction of dormancy. Their material was apparently dead and, it would appear, so is the theory of fatty acid regulation of dormancy.

2. Metabolism and Dormancy

Metabolic manifestations of dormancy include changes in respiratory metabolism, in protein synthesis, in the activity of enzymes and, perhaps, in the production of proteins which counter germination including the ABA-induced α-amylase inhibitor (see above). Membrane responses may also differ. Furthermore, when a treatment such as reduced oxygen availability controls dormancy, then specific changes in respiration can be argued.[53] Generally, however, the environmental controls which regulate dormancy have a wide range of effects on metabolism. Such nonspecificity is also to be expected for chemical treatments and, in this instance, response may often be pharmacological.

Discussions of change in respiratory metabolism with dormancy have centered on evidence for breaking of dormancy by cyanide and other respiratory inhibitors, high oxygen levels, and nitrate and nitrite, the latter two compounds possibly acting as electron acceptors.[128] The suggestion that activity of the pentose phosphate pathway is elevated when dormancy is broken has not been verified, nor are there changes in the activity of key enzymes of this pathway.[127,129] However, control involving other respiratory paths is not excluded. Sodium azide, for example, enhances the alternative path of respiration and breaks dormancy, while dormancy is maintained by SHAM (salicylhydroxamic acid),[129] an inhibitor of the alternative pathway in wild oat. Even so, there is no unanimous support for the involvement of these respiratory paths in dormancy.[130] Also, there appear to be no essential links between *energy charge*, for example, and dormancy in wild oats.[131]

Dormancy, when found in embryos, may also be expressed in other tissues of the cereal grain. For example, for dormant lines of wild oats, where germination was responsive to

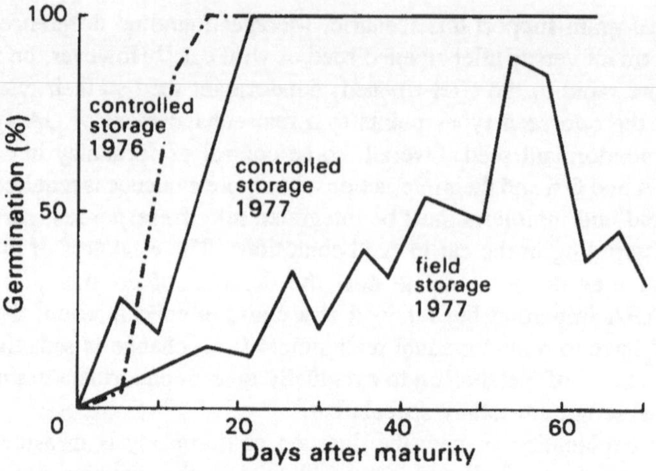

FIGURE 12. Changes in germinability of wheat grain (cv. Lely) following harvest or for the unharvested crop. (Derived from Belderok and Habekotte.[148])

GA, when tested as half-seeds their aleurone response was also dependent on GA, and nondormant lines showed some aleurone GA independence.[101] Clearly, GA-dependence of the aleurone is associated with dormancy; such evidence raises the possibility that understanding aleurone response will lead to an understanding of dormancy.

Another expression of dormancy, the suggestion of change in membrane function, rests on the general observations that solutes leak from seeds during germination; that membrane-active chemicals such as ethanol may break dormancy; and that, in some instances, there are sharp critical temperature responses which are reminiscent of membrane phase changes.[130] However, none of this evidence is yet adequate to establish these responses as important controls of dormancy rather than as changes following from germination.

VI. ENVIRONMENT AND SPROUTING TOLERANCE

A. Dormancy in Field Conditions

How primary dormancy is regulated was discussed above, but its imposition and maintenance requires consideration here. Briefly, the maintenance of primary dormancy reflects the continuing temperatures and rainfall. The positive correlation between low dormancy and high temperature is illustrated in Figure 12 from a study by Belderok and Habekotte in which ears harvested and held at 18°C lost dormancy more rapidly than those held outdoors at lower temperatures.[148] As many others have also reported, it is higher temperatures — even up to 40°C for 2 days — which break primary dormancy. A similar positive relationship between low dormancy and high temperature is seen when the imposition of primary dormancy is considered. Low temperatures most effectively induce high dormancy but now the temperature is most important over the dough ripe stage 5 to 15 days before maturity. Belderok, for example, has provided comprehensive evidence that the temperature sum over a base of 12.5°C for the dough ripe stage of development is inversely correlated with the duration of primary dormancy as shown in Figure 13.[132] Clearly, as he argues, the greater the temperature sum — the higher the daily temperature — then the shorter the dormancy period. He reported similar findings with other cultivars also, although some of these were so dormant that temperature sums were inapplicable.

Year-to-year differences in the imposition and maintenance of dormancy are quite large (Figures 12, 13). A further complicating factor is the inversion of temperature responses

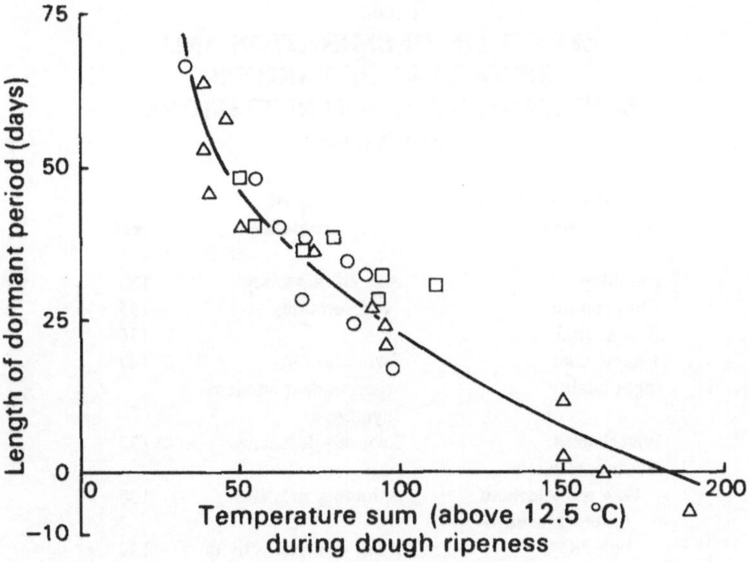

FIGURE 13. Relationship between postharvest dormancy and the temperature sum (base temperature of 12.5°C) during the dough-ripe stage for the cultivar Peko grown in Holland over a number of years. (From Belderok, B., *Eucarpia Proc. Dijon*, p. 111, 1971. With permission.)

when mature grain germinates. When the grain is wet, high temperatures are actually unfavorable for germination (Figure 7); thus, a temperature sum/dormancy index becomes somewhat confounded when used to predict sprouting. In fact, following Belderok's use in the 1960s of an inverse temperature to dormancy relationship to predict sprouting risk, it soon became clear that a more complex set of environmental factors including light, temperature, humidity, and windspeed better defined the length of primary dormancy.[133] Temperature alone, as an indicator, was found to be quite inadequate.[134] Only more recently was the importance recognized of the rate of drying at maturity.[9,104] In fact, there may be no distinction between environment during grain drying and that over the dough ripe period of calculation of a temperature sum. Conditions favorable for rapid drying might even mimic the reduced dormancy of seed developed with large temperature sums, high winds, high insolation, etc. As yet, the only substantial study has been reported by Gale et al.[11] They found that sprouting (germination of isolated grain) was clearly enhanced by allowing rapid drying of the grain from 10 days before to well after maturity. For plants in which the ear and grain dried and matured more slowly (high humidity treatment) the duration of post-maturity dormancy was much greater: 35 days vs. 20 days for slow and fast drying, respectively. The growing temperature used in these experiments (approximately 14°C average) would have favored dormancy, while fast drying acted in the manner of higher temperature to reduce dormancy. Physiological explanations of these drying responses may involve change in grain ABA levels.[82,88] As yet, however, there is no substantial evidence for or against a role for ABA in determining dormancy in cereals, let alone the suggestion that greater loss of ABA occurred with faster grain drying. Indeed, for soybean, detached seed dried rapidly over silica gel (a half loss of water in 1 day) retained most of its ABA, but ABA disappeared with slow drying.[88] For wheat grains, change in ABA with different grain drying rates has not been well documented, but it appears that for both slow and fast drying high ABA levels are maintained, and that it is the intermediate drying rates which give low ABA.[82] As the "fast" drying rate used by Gale et al.[11] was actually rather slow (approximately half loss of water in 10 days vs. 50 days for slow drying), then further examination

Table 7
EFFECT ON GERMINATION AND SPROUTING OF VARIOUS ENVIRONMENTAL AND NUTRITIONAL FACTORS

Environmental factor	Seed response	Ref.
Humidity	Altered dormancy and sprouting	133
Temperature		135
Photoperiod		136
Temperature	Switch to GA-independent amylase synthesis	147
Light quality		
Water stress	Reduced dormancy	137
Nutritional factors		
Low molybdenum (interacts with NO_3)	Sprouting in maize	138
High NO_3	More diastatic activity	139
	More amylase on germination	140
		141
	More sprouting	105
	Enhanced germination vigor	142
Low inorganic PO_4	Enhanced dormancy	143

of grain ABA content might show a parallel with germination. This supposition about ABA change clearly needs examination.

B. Environmental and Edaphic Factors

A range of environmental and edaphic factors enhance or block sprouting in cereals (Table 7). As an indication of response mechanisms, ABA content is increased and, so, germinability might be reduced with high temperature,[70] with reduced potassium supply, and with water stress.[86,87] In wild oat, however, dormancy is reduced by water stress of the parent plant, a treatment which should have inhibited germination were the level of ABA increased.[137] Clearly, it is not yet possible to draw together response and control mechanisms, particularly as many of the studies cited here did not include simultaneous measurements of germination and of ABA.

Often, responses have been sought in the field without adequate control of conditions; however, with proper sprouting assays and/or the use of greenhouse-grown plants, specific conditions favorable to sprouting have now been identified. Response to environmental factors, especially to temperature, is discussed elsewhere. Considering nutrition, the most compelling findings are those of Tanner for corn.[138] By utilizing acid soils (pH 4.3 to 4.7) he showed that the molybdenum content of the corn kernel was critical. Sprouting occurred at levels below 0.5 ppm molybdenum, and this response could be blocked by molybdenum application to the cob. Since molybdenum is important in the enzymic reduction of nitrate and since nitrate stimulates germination, Tanner considered the real lesion was a too high level of nitrate and, as predicted, application of nitrogenous fertilizer exacerbated the sprouting problem. Nitrogen fertilization will also produce grain with higher protein and, hence, grain with enhanced capacity to synthesize α-amylase, as is known for wheat and barley (Table 7). In one case, for wheat grain with protein contents between 16.1 and 17.3%, sprouting was also enhanced — the greater the protein, the greater the sprouting.[105] Others have failed to find such differences in sprouting. However, for protein contents ranging from 14 to 15%, Huang and Marston reported increases in capacity to produce α-amylase.[141]

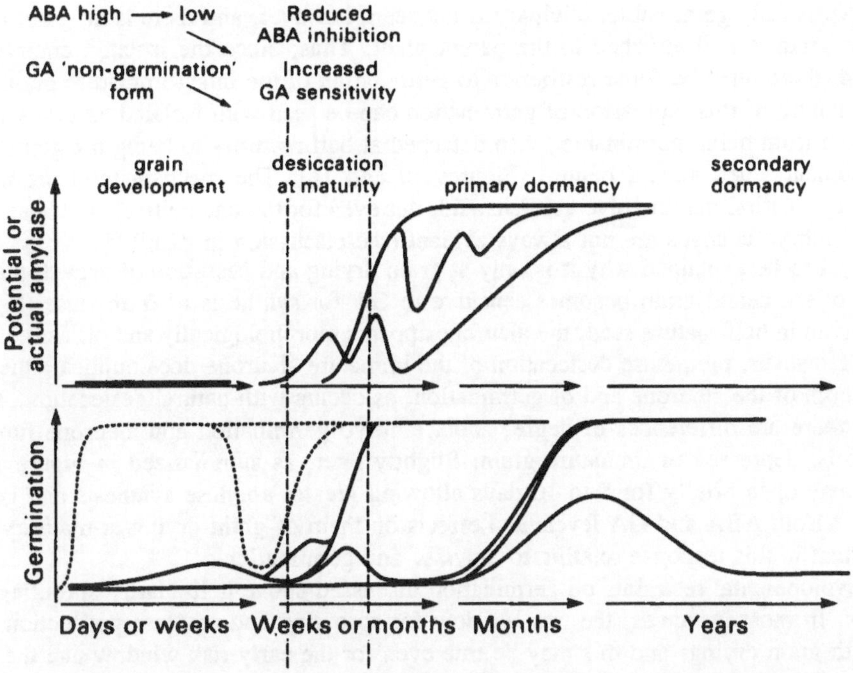

FIGURE 14. Model of the pattern of change in time in cereal grain germinability and capacity to produce α-amylase. Interactions with changes in GA and ABA are also indicated.

It is not known if there is any association between grain protein content and the forms of amylase present. However, Hayter and Riggs suggested that increased diastatic activity in barley reflected an increase in β- and not α-amylase.[139] As β-amylase can synergize the action of α-amylase, then it is possible that some of the assayed increase in "α-amylase" could be an indirect response to β-amylase.[144]

For the future, it is clear that further examination is required of many aspects of the response to fertilizer treatments (Table 7). This is particularly urgent because breeders selecting for high grain protein may be unwittingly producing lines with greater sprouting; certainly germination is enhanced.[142]

Of the other nutrients, phosphate fertilization leads to a reduced dormancy in wild oats.[143] Calcium availability is also of potential importance as there is a specific requirement for calcium for one or a number of the steps in the production of the high isoelectric point groups of amylase isozymes in barley.[145]

VII. AN OVERVIEW OF SPROUTING — ITS TIMING AND SEQUENCE OF RESPONSES

In any discussion of sprouting, it must be shown how time and morphological development of the grain influence the response. To the general pattern of change with time in germination response of cereals (Figure 14), can be added the changes seen for potential and actual α-amylase formation as in Figure 2. Most information is available for wheat, but the same general picture has been noted for barley.[4,82,103]

One important switch during grain development occurs on grain drying and, even when only 5 to 10 days old, the embryo may germinate if first dried. The whole grain is generally not germinable, but capacity for germination may result just as, or slightly before, natural desiccation and cessation of dry matter increase occur. Some days prior to grain maturation, the embryo is fully germinable. Vivipary is not seen, however, and there is no germination while the grain is still attached to the parent plant. Thus, since the isolated embryo can germinate, there must be some restriction to germination of the intact immature grain. The dramatic nature of this repression of germination can be seen with isolated embryos which may switch from being germinable when detached at half maturity to being nongerminable on detachment when almost mature (Figures 10 and 14). The mechanism(s) preventing embryo germination may involve abscisic acid, but even for the one cultivar of wheat, such dramatic embryo changes are not always evident (see discussion in King).[82]

It has yet to be explained why it is only at grain drying and cessation of growth that the aleurone of the cereal grain becomes sensitive to GA for synthesis of α-amylase (Figures 2, 14). Even in half-mature seed, the aleurone appears morphologically and ultrastructually mature.[6] However, premature desiccation of the immature aleurone does duplicate the sensitizing, both of the aleurone and of germination, as occurs with natural desiccation. Thus, although there are differences of degree, both embryo germination and aleurone function appear to be depressed in immature grain. Slightly later, as summarized in Figure 14, a window may open briefly for 5 to 10 days allowing greater amylase synthesis and germination.[11,74] Both ABA and GA levels and effects on them of grain drying at maturity may be important in this response relating to amylase and germination.

The developmental restraints on germination discussed above help clarify sprouting risk in cereals. In most instances, the risk window for sprouting and amylase production only opens with grain drying, and this may be true even for the early risk window and the time of high amylase production of some English winter wheats matured in cool wet conditions (see above and Reference 74). Viewed in the light of these findings, rainfall prior to grain desiccation is of no consequence, the more so as there is no net water uptake prior to this time.[42] As an aside, to examine sprouting at the rather ill-defined stages of "harvest ripe", "mature", or "yellow ripe" is of limited value not only for its imprecision but also because of the importance of earlier response to the rate and timing of grain drying via effects on ability to produce amylase and to germinate. In a few seasons and in some climates, if there are cool wet conditions over the later stages of grain maturation, it may be difficult to identify precisely a time of maturity from measurement of grain water content. Nevertheless, for data collected in Canada over 12 seasons, Dodds and Pelton found ear drying times and rates were relatively consistent.[149] The seasons which may break this generalization are of particular interest as, for example, in the study of Gale et al.,[74] who have now found an "early risk" of high amylase production for English winter wheat crops exposed to cool moist conditions at the time that grain desiccation would normally occur. This elevated production of α-amylase occurred even in the absence of visible sprouting. That grain drying might be critical in the novel early risk window (Gale et al.,[74] Figure 13) is supported by the finding that most amylase activity was found in the grain of basal florets, especially floret 2, and in the central spikelets of the ear. Probably these grains do mature and dry out later and more slowly than the distal grains.

There is a striking contrast to this "early risk" maturity-induced, amylase production in the central grain of an ear to that for grain matured in warmer, drier conditions. Such,mature dry grain has little or no detectable amylase but, when germinated, it is the grain from distal floret positions which produces most amylase and is most germinable.[146] Exposure in-the-ear to simulated rainfall results in much the same pattern but with the further complexity of greater response in tip spikelets; this finding may point to interception of more rainfall at the tip of the ear.

Following its drying at maturity, the grain may become fully germinable and capable of synthesizing α-amylase or become dormant with suppression of both responses. Such primary dormancy may last up to 2 months (Figure 14), and its early onset and continued expression must be maintained if sprouting damage is to be avoided. Of little relevance will be the later imposition of secondary dormancy, although there may be serious consequences for producing seed stocks of wheat and for barley held for malting.

The duration of dormancy is very sensitive to environment and is shortened, particularly, by high temperatures. Temperature response is confused, however, by the switch in response seen when seed is moist rather than dry (Figure 7 vs. Figure 13). Dormancy, too, is dependent on a wider range of environmental variables than temperature alone. The possibility arises, therefore, that rates of grain drying at maturation could integrate all environmental signals and so determine the duration of dormancy. No common response can be suggested for subsequent effects of temperature in determining dormancy of dry, mature grain. This later period may be of less importance, however, in determining susceptibility to sprouting; certainly timing of response requires much more detailed study. The use of controlled environments would obviously help clarify many of these interacting effects of environment and stage of grain development. Field responses, by contrast, may be even too complex for the most complicated multivariate analysis to unravel the important relationships governing sprouting, dormancy, varieties, and their stability across seasons and sites.

REFERENCES

1. **Ashford, A. E. and Gubler, F.,** Mobilization of polysaccharide reserves from endosperm, in *Seed Physiology,* Vol. 2, Murray, D., Ed., Academic Press, Sydney, 1984, 117.
2. **Percival, T.,** *The Wheat Plant,* Duckworth, London, 1921, 463.
3. **Rijven, A. H. G. C. and Cohen, R.,** Distribution of growth and enzyme activity in the developing grain of wheat, *Aust. J. Biol. Sci.,* 14, 552, 1961.
4. **King, R. W.,** Abscisic acid in developing wheat grains and its relationship to grain growth and maturation, *Planta,* 132, 43, 1976.
5. **King, R. W.,** The Physiology of pre-harvest sprouting — a review, in *3rd Int. Symp. Pre-harvest Sprouting in Cereals,* Kruger, J. E. and LaBerge, D., Eds., Westview Press, Boulder, 1983, 11.
6. **Morrison, I. N., Kuo, J., and O'Brien, T. P.,** Histochemistry and fine structure of developing wheat aleurone cells, *Planta,* 123, 105, 1975.
7. **King, R. W.,** Pre-harvest sprouting in wheat: effects of environment, maturity, and genotype, in preparation.
8. **Bingham, J.,** Grain quality from the genetic and plant breeding point of view, *Getreide Mehl,* 18, 61, 1968.
9. **King, R. W. and Gale, M. D.,** Preharvest assessment of potential α-amylase production, *Cereal Res. Commun.,* 8, 157, 1980.
10. **King, R. W., Salminen, S. O., Hill, R. D., and Higgins, T. J. V.,** Abscisic-acid and gibberellin action in developing kernels of triticale (cv. 6A190), *Planta,* 146, 249, 1979.
11. **Gale, M. D., Flintham, J. E., and Arthur, E. D.,** Alpha-amylase production in the late stages of grain development — an early sprouting damage risk period?, in *3rd Int. Symp. Pre-harvest Sprouting in Cereals,* Kruger, J. E. and LaBerge, D., Eds., Westview Press, Boulder, 1983, 29.
12. **Fischnich, O. and Thielebein, M.,** Hinewise für den Mähreschereinsatz, *Mitt. Dtsch. Landwirtsch. Ges.,* 69, 545, 1954.
13. **King, R. W. and Richards, R. A.,** Water uptake and pre-harvest sprouting damage in wheat: ear characteristics, *Aust. J. Agric. Res.,* 35, 327, 1984.
14. **McMaster, G. J. and Derera, N. F.,** Methodology and sample preparation when screening for sprouting damage in cereals, *Cereal Res. Commun.,* 4, 251, 1976.
15. **Kinnell, P. I. A.,** The problem of assessing the erosive power of rainfall from meterological observations, *Soil Sci. Soc. Am. Proc.,* 37, 617, 1973.
16. **Hardy, K. R. and Dingle, A. N.,** Raindrop size distributions in a cold frontal shower, Proc. 8th Weather Radar Conf., Cambridge, Mass., 179, 1960.
17. **King, R. W.,** unpublished results, 1986.
18. **Mares, D. J.,** unpublished results, 1986.
19. **Mares, D. J.,** Investigation of the pre-harvest sprouting damage resistance mechanisms in some Australian white wheats, in *3rd Int. Symp. Pre-harvest Sprouting in Cereals,* Kruger, J. E., and LaBerge, D., Eds., Westview Press, Boulder, 1983, 59.

20. **Pool, M. and Patterson, F. L.,** Moisture relations in soft red winter wheats. XI. Awned versus awnless and waxy versus nonwaxy glumes, *Agron. J.*, 50, 158, 1958.
21. **Brinkman, M. A. and Luk, T. M.,** Relationship of spike nodding angle and kernel brightness under simulated rainfall in barley, *Can. J. Plant Sci.*, 59, 481, 1979.
22. **Akalehiywot, T. and Bewley, J. D.,** Dessication of oat grains during and following germination and its effects upon protein synthesis, *Can. J. Bot.*, 58, 2345, 1980.
23. **Lush, W. M., Groves, R. H., and Kaye, P. E.,** Presowing hydration-dehydration treatments in relation to seed germination in early seedling growth of wheat and ryegrass, *Aust. J. Plant Physiol.*, 8, 409, 1981.
24. **Shaykewich, C. F.,** Proposed method for measuring swelling pressure of seeds prior to germination, *J. Exp. Bot.*, 24, 1056, 1973.
25. **Clarke, J. M.,** Measurement of relative water uptake rates of wheat seeds using agar media, *Can. J. Plant. Sci.*, 60, 1035, 1980.
26. **Owen, P. C.,** The relation of water absorption by wheat seeds to water potential, *J. Exp. Bot.*, 3, 276, 1952.
27. **Leopold, A. C.,** Volumetric components of seed imbibition, *Plant Physiol.*, 73, 677, 1983.
28. **Wellington, P. S.,** Studies on the germination of cereals. II. Factors determining the germination behaviour of wheat grains during maturation, *Ann. Bot. (London)*, 20, 481, 1956.
29. **Murphy, J. B. and Noland, T. L.,** Temperature effects on seed imbibition and leakage mediated by viscosity and membranes, *Plant Physiol.*, 69, 428, 1982.
30. **Kaufmann, M. R. and Ross, K. J.,** Water potential, temperature and kinetin effects on seed germination in soil and solute systems, *Am. J. Bot.*, 57, 413, 1970.
31. **Owen, P. C.,** The relation of germination of wheat to water potential, *J. Exp. Bot.*, 3, 188, 1952.
32. **King, R. W.,** Water uptake and preharvest sprouting damage in wheat:grain characteristics, *Aust. J. Agric. Res.*, 35, 337, 1984.
33. **Butcher, J. and Stenvert, N. L.,** Conditioning studies on Australian wheat. III. The role of the rate of water penetration into the wheat grain, *J. Sci. Food Agric.*, 24, 1077, 1973.
34. **Moss, R.,** Conditioning studies on Australian wheat. III. Morphology of wheat and its relationship to conditioning, *J. Sci. Food Agric.*, 24, 1067, 1973.
35. **Marshall, D. R., Ellison, F. W., and Mares, D. J.,** Effects of grain shape and size on milling yields in wheat, I. Theoretical analysis based on simple geometric models, *Aust. J. Agric. Res.*, 35, 619, 1984.
36. **Woodbury, W. and Wiebe, T. J.,** A possible role for the pericarp in control of germination and dormancy of wheat, in *3rd Int. Symp. Pre-harvest Sprouting in Cereals*, Kruger, J. E. and LaBerge, D., Eds., Westview Press, Boulder, 1983, 51.
37. **Reynolds, T. and MacWilliam, T. C.,** Water uptake and enzymic activity during steeping of barley, *J. Inst. Brew. London*, 72, 1166, 1966.
38. **Briggs, D. E. and MacDonald, J.,** The permeability of the surface layers of cereal grains and implications for tests of abrasion in barley, *J. Inst. Brew. London*, 89, 324, 1983.
39. **Freeman, P. L. and Palmer, C. H.,** The influence of the pericarp and testa on the direct passage of gibberellic acid into the aleurone of normal and abraded barleys, *J. Inst. Brew. London*, 90, 95, 1984.
40. **Hinton, J. J. C.,** Resistance of the testa to entry of water into the wheat kernel, *Cereal Chem.*, 32, 296, 1955.
41. **Collins, E. J.,** The structure of the integumentary system of the barley grain in relation to localized water absorption and semi-permeability, *Ann. Bot. (London)*, 32, 381, 1918.
42. **King, R. W.,** Ear and grain wetting in relation to pre-harvest sprouting, in *4th Int. Symp. Pre-harvest Sprouting in Cereals*, Mares, D. J., Ed., Westview Press, Boulder, 1987, 327.
43. **Mares, D. J.,** Temperature dependence of germinability of wheat (*Triticum aestivum* L.) grain in relation to pre-harvesting sprouting. *Aust. J. Agric. Res.*, 35, 115, 1984.
44. **Corbineau, F., Lenoir, C. Lecat, S., and Côme, D.,** La dormance des semences des cereals et son elimination (suite) 2 Mecanismes mis en jeu, *C. R. Seances Acad. Agric. Fr.*, 70, 717, 1984.
45. **Côme, D. and Corbineau, F.,** La dormance des semences des cereales et son elimination. I. Principles caracteristiques, *C. R. Seances Acad. Agric. Fr.*, 70, 709, 1984.
46. **Marcus, A., Feeley, J., and Volcani, T.,** Protein synthesis in imbibed seed. III. Kinetics of amino acid incorporation, ribosome activation, and polysome formation, *Plant Physiol.*, 41, 1167, 1966.
47. **Chen, S. S. C. and Varner, J. E.,** Metabolism of ^{14}C-maltose in *Avena sativa* seeds during germination, *Plant Physiol.*, 44, 770, 1969.
48. **Bewley, J. D. and Black, M.,** *Physiology and Biochemistry of Seeds in Relation to Germination*, Vol. 1, Springer-Verlag, Berlin, 1978, 306.
49. **Jones, R. L.,** Quantitative and qualitative changes in the endoplasmic reticulum of barley aleurone layers, *Planta*, 150, 70, 1980.
50. **Atzorn, R. and Weiler, E. W.,** The role of endogenous gibberellins in the formation of α-amylase by aleurone layers of germinating barley caryopses, *Planta*, 159, 289, 1983.

51. **Brown, H. T. and Morris, G. H.**, Researches on the germination of some of the *Gramineae*, *J. Chem. Soc.*, 57, 458, 1890.
52. **Groat, J. I. and Briggs, D. E.**, Gibberellins and α-amylase formation in germinating barley, *Phytochemistry*, 8, 1615, 1969.
53. **Hsiao, A. I.**, Mechanisms of dormancy in wild oats (*Avena fatua*), in *4th Int. Symp. Pre-harvest Sprouting in Cereals*, Mares, D. J. Ed., Westview Press, Boulder, 1987, 425.
54. **Mares, D. J.**, Rate and location of production of alpha-amylase in relation to pre-harvest sprouting in wheat, in *4th Int. Symp. Pre-harvest Sprouting in Cereals*, Mares, D. J., Ed., Westview Press, Boulder, 1987, 494.
55. **Gilmour, S. J. and MacMillan, J.**, Effect of inhibition of gibberellin biosynthesis on the induction of α-amylase in embryoless caryopses of *Hordeum vulgare* cv. Himalaya, *Planta*, 162, 89, 1984.
56. **Jacobsen, J. V. and Chandler, P. M.**, Gibberellin and abscisic acid in germinating cereals, in *Plant Hormones and Their Role in Plant Growth and Development*, Davies, P. J., Ed., Nijhoff/Junk, Dordrecht, Netherlands, 1987, 164.
57. **Mikola, L. and Kolehmainen, L.**, Localization and activity of various peptidases in germinating barley, *Planta*, 104, 167, 1972.
58. **Briggs, D. E.**, *Barley*, Chapman and Hall, London, 1978, 612.
59. **Gibbons, G. C.**, Immunohistochemical determination of the transport pathway of α-amylase in germinating barley seeds, *Cereal Res. Commun.*, 8, 87, 1980.
60. **MacGregor, A. W. and Matsuo, R. R.**, Starch degradation in endosperms of barley and wheat kernels during initial stages of germination, *Cereal Chem.*, 59, 210, 1982.
61. **MacGregor, A. W., MacDougall, F. H., Mayer, C., and Daussant, J.**, Changes in level of α-amylase components in barley tissues during germination and early seedling growth, *Plant Physiol.*, 75, 203, 1984.
62. **Ranki, H. and Sopanen, T.**, Secretion of α-amylase by the aleurone layer and the scutellum of germinating barley grain, *Plant Physiol.*, 75, 710, 1984.
63. **Okamoto, K., Kitano, H., and Akazawa, T.**, Biosynthesis and excretion of hydrolases in germinating cereal seeds, *Plant Cell Physiol.*, 21, 201, 1980.
64. **Cornford, C. A., Black, M., and Chapman, J.**, Sensitivity of developing wheat grains to gibberellin and production of α-amylase during grain development and maturation in *4th Int. Symp. Pre-harvest Sprouting in Cereals*, Mares, D. J., Ed., Westview Press, Boulder, 1987, 283.
65. **Palmer, G. H.**, The morphology and physiology of malting barleys, in *Cereals for Food and Beverages*, Inglett, G. E., and Munck, L., Eds., Academic Press, New York, 1980, 301.
66. **Palmer, G. H.**, A reassessment of the pattern of endosperm hydrolysis (modification) in germinating barley, *J. Inst. Brew. London*, 88, 145, 1982.
67. **Pharis, R. P. and King, R. W.**, Gibberellins and reproductive development in seed plants, *Annu. Rev. Plant Physiol.*, 36, 517, 1985.
68. **Gaskin, P., Kirkwood, P. S., Lenton, J. R., MacMillan, J., and Radley, M. E.**, Identification of gibberellins in developing wheat grain, *Agric. Biol. Chem.*, 44, 1589, 1980.
69. **Lenton, J. R. and Gale, M. D.**, Hormonal changes during cereal grain development, in *4th Int. Symp. Pre-harvest Sprouting in Cereals*, Mares, D. J., Ed., Westview Press, Boulder, 1987, 253.
70. **Radley, M.**, The development of wheat grain in relation to endogenous growth substances, *J. Exp. Bot.*, 27, 1009, 1976.
71. **Gaskin, P., Gilmour, S. J., Lenton, J. R., MacMillan, J., and Sponsel, V. M.**, Endogenous gibberellins and kaurenoids identified from developing and germinating barley grain, *J. Plant Growth Regul.*, 2, 229, 1983.
72. **Dathe, W. and Sembdner, G.**, Distribution of gibberellins and abscisic acid in different fruit parts of rye, Gibberellins, *Biochem. Physiol. Pflanzen*, 173, 440, 1978.
73. **Metzger, J. D.**, Role of endogenous plant growth regulators in seed dormancy of *Avena fatua*. II. Gibberellins, *Plant Physiol.*, 73, 791, 1983.
74. **Gale, M. D., Salter, A. M., and Lenton, J. R.**, The induction of germination α-amylase during wheat grain development in unfavourable weather conditions, in *4th Int. Symp. Preharvest Sprouting in Cereals*, Mares, D. J., Ed., Westview Press, Boulder, 1987, 273.
75. **Phinney, B. O.**, Gibberellin A_1 dwarfism and the control of shoot elongation in higher plants, in *The Biosynthesis and Metabolism of Plant Hormones*, S.E.B. Seminar Ser. 23, Oxford University Press, Oxford, 1984, 17.
76. **Wheeler, A. W.**, Changes in growth substance contents during growth of wheat grains, *Ann. Appl. Biol.* 72, 327, 1974.
77. **Black, M. and Naylor, J. M.**, Prevention of the onset of seed dormancy by gibberellic acid, *Nature (London)*, 184, 468, 1959.
78. **Rood, S. B., Pharis, R. P., and Koshioka, M.**, Reversible conjugation of gibberellins *in situ* in maize, *Plant Physiol.*, 73, 340, 1983.

79. **Smith, M. T. and Briggs, D. E.,** Externally applied gibberellic acid and α-amylase formation in grains of barley *(Hordeum distichon), Phytochemistry,* 19, 1025, 1980.
80. **Schopfer, P. and Plachy, C.,** Control of seed germination by abscisic acid. III. Effect on embryo growth potential (minimum turgor pressure) and growth coefficient (cell wall extensibility) in *Brassica napus* L., *Plant Physiol.,* 77, 676, 1985.
81. **Mundy, J.,** Hormonal regulation of α-amylase inhibitor synthesis in germinating barley, *Carlsberg Res. Commun.,* 49, 365, 1984.
82. **King, R. W.,** Abscisic acid in seed development, in *The Physiology and Biochemistry of Seed Development,* Khan, A. A., Ed., Elsevier, Amsterdam, 1982, 157.
83. **King, R. W.,** Abscisic acid synthesis and metabolism in wheat ears, *Aust. J. Plant Physiol.,* 6, 99, 1979.
84. **Walker-Simmons, M. K.,** ABA levels and sensitivity in developing wheat embryos of sprouting resistant and susceptible cultivars, *Plant Physiol.,* 84, 61, 1984.
85. **Yamada, K.,** Endogenous abscisic acid in barley and use of abscisic acid in malting, *Agric. Biol. Chem.,* 49, 429, 1985.
86. **Goldbach, H. and Goldbach, E.,** Abscisic acid translocation and influence of water stress on grain abscisic acid content, *J. Exp. Bot.,* 28, 1342, 1977.
87. **Haeder, H. E. and Beringer, H.,** Influence of potassium nutrition and water stress on the content of abscisic acid in grains and flag leaves of wheat during grain development, *J. Sci. Food Agric.,* 32, 552, 1981.
88. **Ackerson, R. C.,** Abscisic acid and precocious germination in soybeans, *J. Exp. Bot.,* 35, 414, 1984.
89. **Goldbach, H. and Michael, G.,** Abscisic acid content of barley grains during ripening as affected by temperature and variety, *Crop Sci.,* 16, 797, 1976.
90. **Slominski, B., Rejowski, A., and Nowak, J.,** Abscisic acid and gibberellic acid contents in ripening barley seeds, *Physiol. Plant.,* 45, 167, 1979.
91. **Berrie, A. M. M., Buller, D. R., Don, R., and Parker, W.,** Possible role of volatile fatty acids and abscisic acid in the dormancy of oats, *Plant Physiol.,* 63, 758, 1979.
92. **Black, M.,** The role of endogenous hormones in germination and dormancy, *Isr. J. Bot.,* 25, 181, 1980/81.
93. **Triplett, B. A. and Quatrano, R. S.,** Timing, localization and control of wheat germ agglutinin synthesis in developing wheat embryos, *Dev. Biol.* , 91, 491, 1982.
94. **Fong, F., Koehler, D. E., and Smith, J. D.,** Fluridone induction of vivipary during maize seed development, in *3rd Int. Symp. Pre-harvest Sprouting in Cereals,* Kruger, J. E. and LaBerge, D., Eds., Westview Press, Boulder, 1983, 188.
95. **Moore, R., Smith, J. D., and Fong, F.,** Gravitropism in abscisic acid deficient seedlings of *Zea mays, Am. J. Bot.,* 72, 1311, 1985.
96. **Karssen, C. M., Brinkhorst-van der Swan, D. L. C., Breckland, A. E., and Koornneef, M.,** Induction of dormancy during seed development by endogenous abscisic acid: studies on abscisic acid deficient genotypes of *Arabidopsis thaliana* (L.) Heynh., *Planta,* 157, 158, 1983.
97. **Radley, M.,** Role of growth substances in the regulation of germination in developing and ripening wheat grains, *Cereal Res. Commun.,* 8, 131, 1980.
98. **Lenton, J. R.,** unpublished results, 1986.
99. **Jackson, D. I.,** Factors affecting response in the barley-endosperm bioassay for gibberellins, *J. Exp. Bot.,* 22, 613, 1971.
100. **Nicholls, P. B.,** The effect of the environment during grain growth in barley on gibberellin physiology, in *4th Int. Symp. Pre-Harvest Sprouting in Cereals,* Mares, D. J., Ed., Westview Press, Boulder, 1987, 245.
101. **Unadhyaya, M. K., Simpson, G. M., and Naylor, J. M.,** The physiological basis of seed dormancy in *Avena fatua* L., I. Action of the respiratory inhibitors sodium azide and salicylhydroxamic acid, *Physiol. Plant.,* 54, 419, 1982.
102. **Harvey, B. M. R. and Oaks, A.,** The role of gibberellic acid in the hydrolysis of endosperm reserves in *Zea Mays, Planta,* 121, 67, 1974.
103. **Evans, M., Black, M., and Chapman, J.,** Induction of hormone sensitivity by dehydration is one positive role for drying in cereal seed, *Nature (London),* 258, 144, 1976.
104. **Nicholls, P. B.,** Induction of sensitivity to gibberellic acid in developing wheat caryopses: Effect of rate of desiccation, *Aust. J. Plant Physiol.,* 6, 229, 1979.
105. **Morris, C. F. and Paulsen, G. M.,** Pre-harvest sprouting of hard winter wheat as affected by nitrogen nutrition, *Crop Sci.,* 25, 1028, 1985.
106. **Gordon, I. L.,** Germinability, dormancy and grain development, *Cereal Res. Commun.,* 8, 115, 1980.
107. **Gordon, I. L.,** Factor analysis of characters useful in screening wheat for sprouting damage, in *3rd Int. Symp. Pre-harvest Sprouting in Cereals,* Kruger, J. E. and LaBerge, D., Eds., Westview Press, Boulder, 1983, 231.
108. **Meredith, P. and Pomeranz, Y.,** Sprouted grain, *Adv. Cereal Sci. Technol.,* 7, 239, 1985.

109. **Häggquist, M. L., Pettersson, A., and Lijenberg, C.,** Growth inhibitors in oat grains. I. Leakage of inhibiting factors from oat grains during imbibition and the effect of these on germination and growth, *Physiol. Plant.*, 61, 75, 1984.
110. **Harvey, B. L., Rossnagel, B. G., and Muderewich, R. P.,** Sprouting resistance in barley, in *3rd Int. Symp. Pre-harvest Sprouting in Cereals*, Kruger, J. E. and LaBerge, D., Eds., Westview Press, Boulder, 1983, 239.
111. **Miyamoto, T., Tolbert, N. E., and Everson, E. H.,** Germination inhibitors related to dormancy in wheat seeds, *Plant Physiol.*, 36, 739, 1961.
112. **Miyamoto, T. and Everson, E. H.,** Biochemical and physiological studies of wheat seed pigmentation, *Agron. J.*, 50 733, 1958.
113. **Derera, N. F., Bhatt, G. M., and McMaster, G. L.,** On the problem of pre-harvest sprouting in wheat, *Euphytica*, 26, 299, 1977.
114. **Marais, G. F. and Kruis, W. J. G.,** Pre-harvest sprouting — the South African situation for seed dormancy in barley, in *3rd Int. Symp. Pre-harvest Sprouting in Cereals*, Kruger, J. E. and LaBerge, D., Eds., Westview Press, Boulder, 1983, 267.
115. **Belderok, B.,** Studies on the causes of dormancy of wheat, *Nord. Cerealkjemikerforenings Kong. I. Oslo*, Institute for Grain, Flour, and Bread, Wageningen, p. 67, 1960.
116. **Côme, D., Lenoir, C., and Corbineau, F.,** La dormance des céréales et son élimination, *Seed Sci. Technol.*, 12, 629, 1984.
117. **Lenoir, C., Corbineau, F., and Côme, D.,** Role des glumelles dans la dormance des semences d'orge, *Physiol. Veg.*, 21, 633, 1984.
118. **Corbineau, F., Lenoir, C., and Côme, D.,** Problems related to storage and malting of dormant barley seeds, *Proc. 7th World Cereal and Bread Congr.*, 371, 1982.
119. **Durham, V. M. and Wellington, P. S.,** Studies on the germination of cereals, IV. The oxygen requirement for germination of wheat grains during maturation, *Ann. Bot. (London)*, 25, 197, 1961.
120. **Harrington, G. T.,** Forcing the germination of freshly harvested wheat and other cereals, *J. Agric. Res.*, 23, 79, 1923.
121. **Nilsson-Ehle, H.,** Zur Kentnis der mit der Keimungsphysiologie des Weizens in Zusammenhang stehenden inneren Faktoren, Z. *Pflanzenzuecht.*, 3, 153, 1914.
122. **Freed, R. D., Everson, E. H., Ringlund, K., and Gullord, M.,** Seed coat color in wheat and the relationship to seed dormancy at maturity, *Cereal Res. Commun.*, 4, 147, 1976.
123. **DePauw, R. M., McCaig, T. N., and Townely-Smith, T. L.,** Registration of sprouting resistance white-seeded spring wheat germplasm line, *Crop Sci.*, 25, 577, 1985.
124. **Stoy, V. and Olsen, O.-A.,** Inheritance of a factor affecting the response to germination inhibitors in excised wheat embryo, *Cereal Res. Commun.*, 8, 203, 1980.
125. **Walker-Simmons, M. and Sesing, J.,** Development of a sensitive immunoassay for abscisic acid in wheat grain utilising a monoclonal antibody, in *4th Int. Symp. Pre-harvest Sprouting in Cereals*, Mares, D. J., Ed., Westview Press, Boulder, 1987, 591.
126. **Koornneef, M., Jorna, M. L., Brinkenhorst-van der Swan, D. L. C., and Karssen, C. M.,** The isolation of abscisic acid (ABA) deficient mutants by selection of induced revertants in non-germinating gibberellin sensitive lines of *Arabidopsis thaliana*, *Theor. Appl. Genet.*, 61, 385, 1982.
127. **Metzger, J. D. and Sebesta, D. K.,** Role of endogenous growth regulators in seed dormancy of *Avena fatua*. I. Short chain fatty acids, *Plant Physiol.*, 70, 1480, 1982.
128. **Roberts, E. H. and Smith, R. D.,** Dormancy and the pentosephosphate pathway, in *The Physiology and Biochemistry of Seed Dormancy and Germination*, Khan, A. A., Ed., Elsevier-North-Holland, Amsterdam, 385, 1977.
129. **Upadhyaya, M. K., Simpson, G. M., and Naylor, J. M.,** Levels of glucose-6-phosphate and 6-phosphogluconate dehydrogenase in the embryos and endosperms of some dormant and non-dormant lines of *Avena fatua* during germination, *Can. J. Bot.*, 59, 1640, 1981.
130. **Black, M., Butler, J., and Hughes, M.,** Control and development of dormancy in cereals, in *4th Int. Symp. Pre-harvest Sprouting in Cereals*, Mares, D. J., Ed., Westview Press, Boulder, 1987, 379.
131. **Adkins, S. W. and Ross, J. D.,** Adenosine triphosphate and adenylate energy charge in relation to dormancy of wild oat seeds, *Can. J. Bot.*, 61, 3349, 1983.
132. **Belderok, B.,** Influence of pre-harvest weather conditions on the length of the dormant period in wheat, *Eucarpia* Proc., Dijon, France, p. 111, 1971.
133. **Grahl, A. and Schrödter, H.,** Witterung vorder Reife und Keimruhe von Weizen unter besonderer Beruckschtigung der Auswuchsvorhersage, *Seed Sci. Technol.*, 3, 815, 1975.
134. **Lallukka, U.,** The effect of the temperature during the period prior to ripening on sprouting in the ear in rye and wheat varieties grown in Finland, *Cereal Res. Commun.*, 4, 93, 1976.
135. **Gutterman, Y.,** Influences on seed germinability: phenotypic maternal effects during seed maturation, *Isr. J. Bot.* 29, 105, 1980/81.

136. **Takahashi, N.,** Effect of environmental factors during seed formation on pre-harvest sprouting, *Cereal Res. Commun.*, 8, 175, 1980.
137. **Sawhney, R. and Naylor, J. M.,** Dormancy studies in seed of *Avena fatua,* XIII. Influence of drought stress during seed development on duration of seed dormancy, *Can. J. Bot.*, 60, 1016, 1982.
138. **Tanner, P. D.,** A relationship between premature sprouting on the cob and molybdenum and nitrogen status of maize grain, *Plant Soil,* 49, 427, 1978.
139. **Hayter, A. M. and Riggs, T. J.,** Environmental and varietal differences in diastase power and four associated characteristics of spring barley, *J. Agric. Sci.*, 80, 297, 1972.
140. **Ching, T. M. and Rynd, L.,** Developmental differences in embryos of high and low protein wheat seeds during germination, *Plant Physiol.*, 62, 866, 1978.
141. **Huang, G. and Marston, E. V.,** α-Amylase activity and preharvest sprouting damage in Kansas red and white wheat, *J. Agric. Food Chem.*, 28, 509, 1980.
142. **Torres, J. L., Paulsen, G. M.,** Increasing seed protein content enhances seedling emergence and vigor in wheat, *J. Plant Nutr.*, 5, 1133, 1982.
143. **Quick, W. A., Hsiao, A. I., and Jain, J. C.,** Endogenous inorganic phosphate in relation to seed dormancy and germination of wild oats, *Plant Sci. Lett.*, 28, 129, 1982.
144. **Briggs, D. E.,** Gel-diffusion method for the assay of α-amylase, *J. Inst. Brew. London,* 68, 27, 1962.
145. **Deikman, J. and Jones, R. L.,** Regulation of the accumulation of mRNA for α-amylase isoenzymes in barley aleurone, *Plant Physiol.*, 80, 672, 1986.
146. **King, R. W. and Chadim, H.,** Ear wetting and pre-harvest sprouting of wheat, in *3rd Int. Symp. Pre-harvest Sprouting in Cereals,* Kruger, J. E. and LaBerge, D., Eds., Westview Press, Boulder, 1983, 36.
147. **Nicholls, P. B.,** Development of responsiveness to gibberellic acid in the aleurone layer of immature cereal caryopses: effect of temperature, *Aust. J. Plant Physiol.*, 7, 65, 1980.
148. **Belderok, B. and Habekotte, A.,** Induction of secondary dormancy in wheat crops by low temperatures and high relative humidities in the field, *Cereal Res. Commun.*, 8, 167, 1980.
149. **Dodds, M. E. and Pelton, W. L.,** Effect of weather factors on the kernel moisture of a standing crop of wheat, *Agron. J.*, 59, 181, 1967.

Chapter 4

BIOCHEMISTRY OF PREHARVEST SPROUTING IN CEREALS

J. E. Kruger

TABLE OF CONTENTS

I. INTRODUCTION

In examining the biochemistry of preharvest sprouting, one can take a purely plant physiological point of view and try to examine all the physiological events that are occurring during germination. This would then be a voluminous chapter with much regurgitation of information already found in many excellent treatises.[1-4] Instead, this chapter will focus on the causative agents and the biochemical changes present in the germinated cereal grains which are believed responsible for the deleterious effects on end-product quality.

The most important biochemical components that form during germination and which affect quality are enzymes; these will be discussed in the first section. Of the enzymes, α-amylases are usually considered the most important and, correspondingly, more attention will be devoted to them. A number of other enzyme systems, such as proteolytic enzymes, polyphenol oxidases, etc., could also be important depending on the type of product that the cereal is intended for; these are also covered in the first section. The emphasis will be on wheat, because this cereal in particular, along with rye and triticale, suffers the most due to excess enzymes and, because of this, is the source of the most information. Enzymes from other cereals will be discussed in terms of their similarities and differences.

In situations where sprout damage has been extensive, it is possible that the cereal may suffer direct damage by alteration of the biochemical components which are responsible for the quality of a particular end product. For example, storage proteins have a very important role in determining breadmaking quality. Proteolytic degradation *in situ* may render such proteins useless for this purpose. The second section, therefore, will discuss the mobilization of particular endosperm reserves which may be important from a quality standpoint.

The final section of this chapter will look specifically at typical end-product processes, particularly for wheat-based foods, and attempt to provide some indication of the damage that can result from different degrees of sprouting. Pallatiatives will be given where they exist.

II. ENZYMES IN GERMINATING CEREALS

Certain germinative enzymes, because of their specificities, are known to have extremely deleterious effects on quality. It is important, therefore, to know as much as possible about their chemical and physical properties. Such a full understanding is of particular assistance to researchers who are entering the area of preharvest sprouting and trying to solve the problem by attempting to destroy or alter such enzymes in order to prevent their action, as well as to processors of end products who must deal with excess enzymes. This section will concentrate on those properties believed particularly relevant to understanding the role of such enzymes in degradation of their substrates, either *in situ* or in end-product processes.

A. Carbohydrate-Degrading Enzymes

The carbohydrases of germinating cereals include α- and β-amylases, debranching enzymes, cellulases, and glucanases, as well as many glucosidases. Of these hydrolases, the starch-degrading enzyme α-amylase from wheat is known to have disastrous effects in breadmaking processes and the greatest amount of attention has been devoted to studying its properties. β-Amylase has a synergistic role in this breakdown and much is also known about it. On the other hand, more research is needed to elucidate whether some of the other carbohydrases, i.e., debranching enzymes, may also have subtle roles and consequently be of greater importance than is believed at present.

1. α-Amylase

This enzyme cleaves the α-(1→4) D-glucosidic linkages in starch components. The cleavage is restricted by terminal or α-(1→6) D-interchain glucosidic linkages. The complete mechanisms regarding the breakdown of the high molecular weight polymers, amylose and amylopectin, remain to be established.

a. Methods of Determining

Methods for determination of α-amylase fit into a number of categories, including viscometric, dyed-substrate, turbidometric, fluorometric, colorimetric, gel-diffusion, and reducing sugar assays. Many of these methods will be covered in the chapter dealing with assessment of sprouting damage. There are a number of criteria that must be considered when selecting a method for a particular application, including rapidity, simplicity, cost, sensitivity, specificity, and acceptability. Criteria for assessing sprout damage of a large number of samples might include rapidity, simplicity, and cost. Methods such as the Hagberg falling number,[5] gel diffusion, or dyed-substrate might be ideally suited.[6-10] On the other hand, for research studies on α-amylase levels, sensitivity and specificity are the prime prerequisites. In that case, viscometric,[11] turbidometric, or colorimetric assays with iodine are more suitable.[12-13] For information on the α-amylase level of a milled flour and how it might interact with starch in a breadmaking process, the amylograph method (AACC Method 22-10) is preferred. Reducing sugar methods, with starch as substrate, are sometimes used for assaying the combined action of α- and β-amylase. If β-limit dextrin is used as substrate, α-amylase alone can be determined.[14] The gasing power method is used to provide information on the combined interaction of α- and β-information, damaged starch, and yeast in a rising dough.

b. Anatomical Location and Changes in Levels of the Enzyme during Germination

The formation, levels, and anatomical location of the enzyme α-amylase in the germinating grain has been adequately covered in the section on plant physiology. This section will deal, therefore, with the practical consideration of how such enzyme formation and location may relate to enzyme tests and to distribution of the enzyme in mill fractions.

The amount of α-amylase in cereals depends on the resting stage of the caryopses. In the developing cereal grains, the enzyme level increases following flowering, primarily in the pericarp and testa, and then decreases to a low level as the kernel finally matures.[15] Surprisingly, the levels in immature kernels will decrease if harvested and bench-dried. If frozen, however, they will retain their original levels.[15a] Any evaluation of preharvest sprout damage should be certain, therefore, that increased levels of enzyme are actually due to field sprouting and not to immaturity or frost damage of the sample.

In the resting stage, levels of α-amylase in the cereal grain are quite low but appear to be cultivar dependent. There is an added complication in evaluating enzyme levels in apparently resting cultivars in that incipient sprouting may appear and increase enzyme levels without any apparent visual sprouting. Again, certain cultivars are prone to this phenomenon.

Upon germination, there is the progressive formation of enzyme, initially detectable adjacent to the scutellum and, with later germination, adjacent to the aleurone layer. From a practical point of view, this means that considerably more enzyme ends up in the outer layers of the cereal. Consequently, upon milling, a substantial amount of the enzyme present in a cereal kernel may be removed. The amount removed in the bran is variable ranging from 25 to 74%.[16] Factors responsible for such variability need to be elucidated. One likely explanation could be the degree of sprouting. With severe sprouting, there is greater penetration of enzyme into the endosperm and less would be removed by milling. This variation in distribution of enzyme between bran and endosperm influences the relationship between the falling number, carried out on ground whole wheat, and the amylograph, carried out on the flour.[16] This indicates that caution should be used when trying to ascertain the suitability of a flour for breadmaking by carrying out a falling number determination on the whole grain.

The uneven distribution of enzyme in the caryopses of parcels of wheat is reflected in the distribution of enzymes in mill fractions as shown in Figure 1.[17] Particularly in the later middlings, concomitant with increasing bran contamination, there is a greater amount of α-amylase per gram. Surprisingly, bran flour by itself does not have an abnormally high level on a per gram basis; this could indicate the presence of inhibitors in this tissue. It might be expected that a solution to utilization of sprouted wheat would be to eliminate the later middling stream. As shown in Figure 1 (bottom), the total amount of enzyme removed would not be substantial because of the small amount of fifth and sixth middling flour compared to the first, second, and third middlings.

Recently, two research groups have tried pearling of wheat prior to milling.[18-19] A substantial amount of the α-amylase resident in the branny layer was removed by pearling. Whether this is a practical or economically feasible step to include in commercial milling needs to be established.

c. *Multiple Forms of the Enzyme*

From the first observations over 15 years ago that two multiple forms of α-amylase exist in cereals,[20-21] there has been a continual increase in the number of isozymes that have been detected. This has been largely due to the advent of modern protein isolation techniques, such as polyacrylamide gel electrophoresis and isoelectric focusing. For example, up to 22 separate α-amylase components have been detected in wheat.[22] A great deal of the interest in isozymes has been by plant geneticists and also from researchers trying to use isozyme patterns as markers in varietal identification. Caution should be exercised in evaluating the number of true isozymes present in a cereal, however, as certain multiple forms may be artifacts due to the isolation procedure or by complexing with inhibitors.[23-24] What is known with certainty is that there are two major groups of isozymes present in cereals with mul-

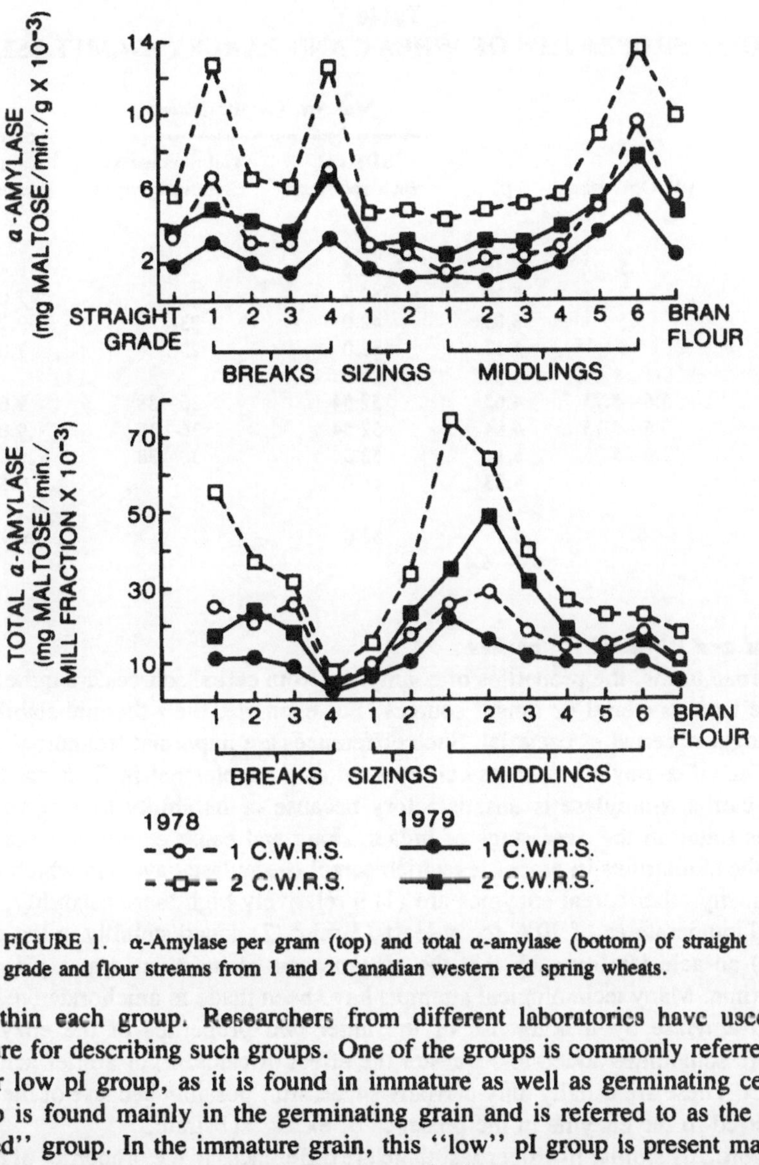

FIGURE 1. α-Amylase per gram (top) and total α-amylase (bottom) of straight grade and flour streams from 1 and 2 Canadian western red spring wheats.

tiplicity within each group. Researchers from different laboratories have used different nomenclature for describing such groups. One of the groups is commonly referred to as the "green" or low pI group, as it is found in immature as well as germinating cereals. The other group is found mainly in the germinating grain and is referred to as the high pI or "germinated" group. In the immature grain, this "low" pI group is present mainly in the pericarp tissue, whereas in the germinating grain, both groups appear to be associated with the aleurone layer and or scutellum. In the germinating grain, the relative proportions of the two groups can vary depending upon the extent of germination.[25] Early germination favors preferential production of "green" α-amylase by the embryo whereas later production by the aleurone layer favors the "germinated" group.

From the above, it is now apparent that the α-amylase systems present in cereals is quite complex. Our information on the individual isozyme continues to grow and, as indicated in the next section, we know a great deal about their properties. The question that remains to be established is the importance of this complexity to the preharvest sprouting problem. If, for example, some of the individual α-amylase isoenzymes work in a synergist or stepwise manner to degrade their natural starch substrate, it would be important to determine which isoenzymes were the earliest in this chain and work towards lowering or eliminating such components by genetic or other means. If, on the other hand, the individual isoenzymes have the same basic action patterns, the total enzymic activity, rather than complexity of pattern, may be the most overriding factor in affecting cereal quality. One study to be presented later suggests the latter is the case.

Table 1
SOME PROPERTIES OF WHEAT AND BARLEY α-AMYLASES

Source	pH Optimum	pI	Mol. wt. (× 10^3 daltons) SDS gel electrophoresis	Mol. wt. (× 10^3 daltons) Gel permeation chromatography	Activation energy (kcal/mol)	Ref.
Malted wheat						
I	5.5	6.16	42.5		8.47	28
II	5.7	6.20	42.2		7.01	
III	5.5	6.05	42.0	23	9.28	
IV	5.5	6.17	42.0	23	7.01	
Immature wheat						
α-1	3.6—5.75	4.65	52.54	36—38	9.66	27
α-2	3.6—5.75	4.84	52.54	36—38	9.04	
α-3	3.6—5.75	5.11	52.54	36—38	8.82	
Immature barley	5.5	4.83	46.0		9.5	26
Malted barley						
I	5.5	5.1	52.0		9.7	108
II		6.2—6.4				

d. Chemical and Physical Properties

In very broad terms, the properties of α-amylase from cereal sources are quite similar but unlike those from bacterial or fungal sources. For example, their thermal stabilities are in the order fungal < cereal < bacterial. Such differences are important technologically. Thus, cereal and fungal α-amylase sources can be used as a supplement in flour used for breadmaking. Bacterial α-amylase is unsatisfactory because of its ability to survive the higher temperatures found in the oven stage of breadmaking and cause excessive dextrinization.

Some of the similarities in properties which cereal α-amylase have and which distinguish them from many other cereal enzymes are (1) a relatively high thermostability, being able to withstand temperatures of 70°C for at least 15 min; (2) added stability in the presence of calcium; (3) an acid lability such that the enzyme can be inactivated at a pH of 3.6 in a fairly short time. Many technological attempts have been made to ameliorate the deleterious effects of α-amylase by making use of the latter two properties of the enzyme. Thus, chemicals are sometimes added to sequester the effect of calcium, or a brief acid treatment is carried out. These are usually only partially successful, possibly because of the stabilizing effect conferred to the enzyme in the presence of excess substrate.

Although broadly similar in properties, there are differences in the properties of α-amylases from different cereals and even individual isozymes and groups of isoenzymes within a cereal. Thus for both wheat and barley, the "green" or low pI α-amylases are more acid stable but heat labile than their "germinated" or high pI components.[26-27]

Some of the fine differences in properties of the individual components from the wheat and barley system are shown in Table 1. Differences in pI values can be attributed to differences which have been found in amino acid composition.[28] It is difficult to believe that such small differences in properties that have been found within a group are required for separate mechanistic roles in breaking down starch. On the other hand, the "green" and "germinated" groups have differences in properties which suggest that they could be operative under different physiological conditions or have separate functional roles.

2. β-Amylase

β-Amylases cleave alternate α-(1→4) D-glucosidic linkages from the nonreducing end of starch substrates resulting in the production of maltose. If amylose is the substrate, even-length chains result in maltose and odd-length chains in maltose and a single D-glucose.

Branched chain substrates such as amylopectin will be degraded to maltose and a β-limit dextrin.

a. Methods of Determining

Determination of β-amylase in cereals is difficult because of interference by α-amylase. In many cases, the combined action of both enzymes is determined or, in the case of sound cereals, it is assumed that α-amylase activity is minimal.

Generally, reducing sugar methods are used with soluble starch as the substrate and measurement of the liberated sugars is carried out with 3,5-dinitrosalicylic acid or neocuproin.[29-30] Some attempt can be made to inactivate the α-amylase preferentially, such as by acid treatment pH at 3.4, but this treatment is not entirely selective and some inactivation of β-amylase can occur. A specific method for β-amylase in cereals has been reported using *p*-nitrophenyl oligosaccharides.[31-32] Only part of the β-amylase is "free" and reductants such as cysteine or a proteolytic enzyme such as papain are required to liberate the "bound" form of the enzyme if total β-amylase is to be measured.

b. Anatomical Location and Changes in Levels of the Enzyme during Germination

β-Amylase is quite different from α-amylase, not only in the tissues in which it is found but in the mechanism by which levels change upon germination. To understand the full picture, we must follow the enzyme throughout kernel development and maturation. In the developing kernel, levels of β-amylase progressively increase in the endosperm, as contrasted with α-amylase which is produced in the pericarp.[15] There is, in addition, a β-amylase component in the pericarp which disappears with kernel maturation. The levels of the endosperm β-amylase components also begin to disappear with final maturation, but not entirely. The enzyme is not being destroyed, however, but bound into the endosperm. Upon germination, there is a several-fold increase in the levels of the enzyme in the endosperm. This increase is not due to the *de novo* synthesis, but to the liberation of the bound enzyme by a mechanism not completely understood but presumably due to a proteolytic or disulfide reductase enzyme mediated release of the enzyme.[33-34]

The situation may be different in rice as Okamota and Akazawa have detected *de novo* synthesis in the scutellum during early germination.[35] With more advanced germination, increases in enzyme levels were by liberation of the "bound" endosperm form of β-amylase.

c. Multiple Forms of the Enzyme

The multiplicity of enzyme forms of cereal β-amylase is very complex, and only certain aspects of this complexity will be examined here. First of all, the number of forms of the enzyme that have been found depends on a number of factors, including the genetic background of the cereal, resolving power of the protein-separating technique, and aggregation of β-amylase components with themselves or to other proteins. With the wheat system, early work detected three major and two minor isoenzymes using ion-exchange chromatography in mature wheat.[36] Gel electrophoresis indicated that two main forms and an additional four minor components were present.[15] The free and bound components were identical in chromatographic properties.[37] Using isoelectric focusing, it was found that the β-amylase present in sound grain was converted to components with higher pI values.[38] This conversion could be simulated by papain treatment.

In barley, much more research has been carried out on the enzyme because of its technological importance in the malting process. The situation appears even more complex. The enzyme is known to be bound to the insoluble hordein proteins and to a soluble Z protein as well.[39] Recent research by LaBerge and Marchylo has indicated that in vitro proteolysis of mature barley β-amylase can effect changes in pI which simulate changes occuring *in situ* as a result of natural germination.[40]

d. Chemical and Physical Properties

In general, wheat and barley β-amylase have many similar properties such as sedimentation coefficient, diffusion constants, partial specific volumes, and sulfhydryl content.[36,41,42] The enzyme is similar to α-amylase in that the pH optimum averages around 5.5, and pI values range from 4 to 6. There are a number of properties, however, which are quite different and which are used to advantage in purification or in devising enzyme assays specific for α-amylase. Thus, β-amylase has greater thermolability. In the purification of α-amylase, heat treatment at 70°C for 15 min is routinely used to selectively inactivate β-amylase. The enzyme is also more acid stable, does not require calcium, and contains sulfhydryl groups essential for catalytic activity. Suflhydryl reagents, such as *p*-chloromercuribenzoate or heavy metals, are used sometimes to selectively inactivate the enzyme in α-amylase assays.

3. Debranching Enzymes

Very little attention has been given in the past to the possible role of debranching enzymes in the preharvest damage problem, although their importance in barley malting is well known. The enzyme can hydrolyze α-(1→6) D-glucosidic linkages in amylopectin, amylopectin β-limit dextrin, and α-dextrins, and this suggests that it certainly could be important. Maeda et al.[43] have reported that although intact starch granules from barley could not be degraded by the enzyme, it did accelerate the α-amylase catalyzed breakdown of the granules. The enzyme is present in a number of cereals in both the ungerminated and germinated form.

a. Methods of Determining

The finding that pullulan is a suitable substrate for cereal debranching enzymes has facilitated development of assays based on measurement of the reaction product, maltotriose, or decrease in viscosity of the substrate.[44-46] A radiochemical method has also been developed using C^{14}-pullulan.[47]

b. Anatomical Location and Changes in Levels of the Enzyme during Germination

In developing wheat, the enzyme has been detected as early as 11 days past anthesis. It is located in the endosperm and persists through to maturity.[48] Mature kernels from different cereals all contain the enzyme in varying amounts. Yamada has found that the levels are in the order oats > barley > wheat > rye.[49] Upon germination, levels of the enzyme again increase due to gibberellic acid-controlled synthesis of the enzyme.[50] Unlike α-amylase, many aspects of the formation of the enzyme need to be elaborated on, such as the relative amounts of the enzyme synthesized in different tissues and whether part of the enzyme may be formed by release of latent, preexisting enzyme.

c. Multiple Forms of the Enzyme

In wheat, two forms of the enzyme have been detected,[48] whereas in barley only one form appears to exist.[51]

d. Chemical and Physical Properties

The chemical and physical properties of the enzyme, particularly with barley, have been studied in some detail. It is known that the pH optima are generally around 5.5,[49,52,53] which is similar to that of the cereal α-amylases. It is unlike α-amylase, however, in not requiring calcium for stability and unlike β-amylase in that sulfhydryl groups are not required for catalytic activity.[48]

4. Other Carbohydrases

There are a number of carbohydrases other than those discussed above that increase upon germination and could be important in affecting quality. Unfortunately, not enough research has been carried out to establish whether this is the case. Such enzymes should be isolated and their role in typical end-product processes evaluated. The enzymes of possible importance include β-glucanases, arabinoxylanases, cellulases, and glycosidases. For example, the endogenous substrate for arabinoxylanases are the water-soluble and water-insoluble pentosans, which are very important in wheat as they affect the water absorption and mixing time of a dough.

B. Protein-Degrading Enzymes

Most researchers feel that proteolytic enzymes formed due to preharvest sprouting are of secondary importance compared to α-amylase in affecting end-product quality. This is undoubtedly so, but no concrete evidence has been presented to indicate the precise extent to which proteolytic enzymes may affect quality. There are a number of reasons for this. First of all, the proteolytic enzyme system is very complex, consisting of endopeptidases, carboxypeptidases, aminopeptidases, and various other peptidases. Secondly, it has been very difficult to purify these enzymes, particularly the endopeptidases, because of their ready association with other proteins and their lability when pure. As a consequence, the effects of adding pure enzymes to test their relative importance in breadmaking, for example, has not been carried out.

In this section, the discussion of proteolytic enzymes will not be exhaustive, but will focus on the specific enzymes which are believed responsible for storage protein hydrolysis. Criteria of importance are the specificity of the enzyme, proximity to substrate, and high enzymic activity. The main enzymes which meet this criteria are the acid endopeptidases and carboxypeptidases. A number of other proteolytic enzymes seem to have lesser or ill-defined roles in the mobilization of endosperm reserves. Undoubtedly, some of them are involved in transport mechanisms.[54,55]

1. Endoproteolytic Enzymes

a. Methods of Determining

The most common method of determining endoproteolytic activity in cereals is based upon the liberation of trichloracetic soluble products from substrates such as hemoglobin or casein. Unfortunately, competitive degradation reactions of such substrates by high levels of the exopeptidase, carboxypeptidase, are such in cereals that they give the appearance of an endopeptidase.[56-58] A preferable substrate to use is azocasein, which appears less sensitive to peptidase action. Some assays have been developed for wheat using the natural gluten substrates. Thus, Kruger developed a method based on the change in stretching properties of gluten after incubation with proteolytic enzymes.[59] Typical rheological tools, such as the farinograph, could also be used to measure changes in gluten viscoelasticity. In another assay by Finlay,[60] sulfanilamide-azogluten was used as substrate and found to be very sensitive for measurement of endoproteases.

b. Anatomical Location and Changes in Levels of the Enzyme during Germination

Endoproteolytic activity, as measured with azocasein, increases during early kernel development in wheat and finally decreases as maturation approaches.[61] Upon germination, the levels increase about eight-fold. Even after 4 days of germination, a portion of the proteolytic activity remains insolubilized.

The location of the enzymes in the early developmental stages of the wheat kernel are in the pericarp, as well as in the seed coat and aleurone. Such enzymes decrease with maturation at the same time as an endoproteolytic enzyme present in the embryo develops, such that at full maturity, the bulk of the enzymic activity is found in this tissue. Enzyme levels also increase in the endosperm, such that at full maturity, one quarter of the enzyme is found in this tissue.[62] Evidently, mechanisms such as the presence of inhibitors, compartmentalization, low moisture, and insolubility of substrate must prevent their action in the mature kernel. The endoproteolytic activity of the barley system appears initially to be synthesized in the aleurone, in much the same way as α-amylase, and is eventually secreted into the endosperm. In wheat and barley, increases in endopeptidase of the germinating kernel are the result of hormonally induced *de novo* synthesis.[63,64]

c. Multiple Forms of the Enzyme

Various numbers of endoproteolytic enzymes (depending on the separatory technique) have been reported from different laboratories. For example, in barley, from two to five enzymes seem to be present.[65,66] In wheat, some authors have reported two enzymes.[67,68] By starch gel electrophoresis at pH 3.4, four groups of proteinases were found.[69] Artifactual proteases, resulting from association of the enzymes with their natural substrates, should be viewed as a distinct possibility in this latter study.

d. Chemical and Physical Properties

Although there have been a number of attempts to purify the endoproteolytic enzymes present in germinating wheat to homogeneity, these have generally met with limited success for reasons discussed previously. Because of this, the properties of pure enzymes still remain to be elucidated.

There have been some properties which have been determined on semipurified preparations and are useful, nevertheless, for providing guidelines for specific purposes such as in devising assays or preferential inactivation of the enzyme from other ones. The pH optimum of the enzyme can vary depending on the substrate and buffer. The soluble form of the enzyme has a pH optimum of 3.8, and the insoluble form 4.4, with hemoglobin as substrate.[67,70] A pH of 6.0 is used in assays with azocasein as substrate by Kruger,[61] but this is due to insolubility of the substrate below this pH rather than a true pH optimum. With edesten as substrate, a pH optimum of 4.1 has been found by Mountfield.[71,72]

The low pH optimum reported above would indicate that under normal breadmaking conditions, the enzyme would not be highly operative.

Endoproteolytic enzymes appear to be sulfhydryl enzymes in that many but not all sulfhydryl blocking agents affect their enzymic activity.[68,70] At least part of the enzyme system is fairly heat labile being inactivated at 40 to 55°C.[71,72]

2. Acid Carboxypeptidases

a. Methods of Determining

These enzymes break down proteins and peptides from the C-terminal amino acid end. One of the most common methods of determining carboxypeptidase activity is to use peptides as substrates with a carbobenzoxymoiety at the N-terminal end of the peptide chain and measure the amino acids liberated. Visuri et al. have devised such an assay with *N*-carboxy-L-phenylalanyl-L-alanine as substrate.[73] The alanine liberated can be determined by an automated procedure on a Technicon Auto Analyzer.[56] Another automated procedure is to measure the dialyzable amino acids liberated from hemoglobin by coupling with fluorescamine.[74]

b. Anatomical Location and Changes in Levels of the Enzyme during Germination

During kernel growth and maturation of wheat, there is a carboxypeptidase system present in the outer layers of the kernel at earlier stages of growth; it subsequently disappears as the kernel matures. In the endosperm, on the other hand, a carboxypeptidase is present which increases throughout the entire development of the kernel.[75] Upon germination, levels of carboxypeptidase increase about three-fold, and this increase appears to be due to removal of inhibitors present in the wheat.[56] The enzyme is concentrated in the endosperm largely adjacent to the scutellar epithelium of the kernel.

When examining other cereals, it appears that carboxypeptidase activity is high in temperate zone cereals such as wheat, barley, oats, rye, and triticale, but lower in cereals such as maize, sorghum or rice, originating in Central America or Asia. The levels of endopeptidase, however, are the reverse.[76]

c. *Multiple Forms of the Enzyme*

There appear to be at least five carboxypeptidases in cereals.[77-79] In wheat, two main carboxypeptidases have been separated by chromatography and purified.[57]

d. *Chemical and Physical Properties*

The carboxypeptidases from wheat and barley appear similar.[57,73] In wheat, studies on the enzyme have shown that its properties and specificity make it a potentially important enzyme in the mobilization of protein reserves. The enzyme is able to rapidly liberate amino acids including proline. This is important since proline can account for around 20% of the prolamines present in wheat.

The enzyme is a serine-type proteolytic enzyme and has an optimum pH of 5.5 for hydrolysis of peptides. Hydrolysis of a soluble gluten fraction was most effective at pH 4.2. Enzymic activity was unaffected by salts such as $CaCl_2$, and sulfhydryl blocking agents also had minimal inhibitory effects. The enzyme was labile at both acid (less than 4) and alkaline pH (greater than 7). Although fairly stable at 45°C, it was rapidly inactivated at 70°C.

3. *Other Proteolytic Enzymes*

There are a myriad of other peptidases present in cereal, and most of them are presumably important in the mobilization of storage proteins. At present, it is doubtful that they have a direct role in affecting end-product quality. For example, enzymes such as the aminopeptidases have a very alkaline pH optima. Furthermore, a large number of them, such as the enzyme-hydrolyzing benzoyl-DL-arginyl *p*-nitroanilide,[80] hydrolyze only di- and tripeptides. As most of them remain to be purified to homogeneity, little mention can be made about them at present.

C. Other Germinative Enzyme Systems

The enzymes to be described below increase in activity during germination, but more research should be carried out to elaborate on their potential importance in affecting quality.

1. *Lipases*

Lipases are known to increase upon germination, with a considerable proportion located in the plumule.[81,82] Tavener and Laidman found that the largest amount of activity was present in the endosperm.[83] One area in which such enzymes would be important is in the postharvest storage of sprouted wheat under elevated moisture and temperature conditions where elevated levels of free fatty acids would occur.

2. *Phenoloxidases, Peroxidases, and Catalases*

Taneja and co-workers, as well as Kruger, have shown that monophenol and polyphenol enzymes increase substantially in wheat upon germination.[84-87] In the latter studies, there were increases of up to 33-fold, and at least 12 isoenzymes were detected with varying specificities towards different substrates. The gray crumb color that results when sprouted wheat is used in breadmaking could well be due to this enzyme complex.

Both peroxidases and catalases from wheat also increase during germination.[88,89] The role of such enzymes is believed to be in the oxidative reactions which affect the rheological property of dough and the destruction of pigments.

III. MOBILIZATION OF ENDOSPERM RESERVES

The mobilization of endosperm reserves can be examined in a number of ways. One approach is to use a microscope (light, scanning electron, or fluorescent) and follow the gross morphological changes that occur with time of germination. This provides an insight into macroscopic changes occurring in the endosperm such as the manner in which starch granules are attacked by α-amylase *in situ*. A second approach is to extract the biochemical components of the grain at different stages of germination and examine the changes in chemical composition of the biochemical reserves as a result of enzymic action. A third approach is to use model systems consisting of purified enzymes and substrate in which it is assumed that similar processes are occurring naturally in the plant. All three approaches provide useful but different information regarding the mobilization of endosperm reserves.

The discussion to follow will be restricted to the starch and protein reserves which comprise the bulk of the endosperm. Both are known to be highly important in determining bread-making quality.

A. Starch Reserves

1. Microscopic Studies

A number of researchers have examined the manner in which starch granules have been attacked by amylases either *in situ* or in solution using the scanning electron microscope (SEM).[90-96] Typical of the damage is the presence of holes in the granule and extensive damage around the equatorial groove. With large granules, there appears to be a preferential hydrolysis of the interior of the granule rather than the surface. The inner rings of the interior of the granule are often visible by SEM. Large and small granules differ in their resistance to hydrolysis. In wheat, the small granules appear less susceptible to enzymic attack, whereas the reverse is found in barley.[96,97] Furthermore, surfaces of small granules often appear rough after hydrolysis, indicating surface as well as interior hydrolysis. A particularly noteworthy study is the one carried out by MacGregor and Matsuo in 1982, since actual field-sprouted samples of durum wheat, rather than laboratory-germinated samples, were used.[96] In that study, the anatomical locations of damage in sound, sprung, and 5-day germinated durum wheat were examined. Typical SEM photomicrographs are shown in Figure 2. Starch damage was minimal in sound kernels with very slight degradation adjacent to the embryo (Figure 2A). In the sprung kernels of wheat (Figure 2B), damage is more apparent, particularly at the crease area adjacent to the embryo, but has not advanced as far as the dorsal region. In the sprouted kernel, the degradation has proceeded along the entire face of the endosperm adjacent to the embryo and was quite extensive. Figure 2D is particularly noteworthy in that typical damage to starch granules is quite apparent.

2. Extraction and Analyses of Starch Components at Different Stages of Germination

There have been very few studies related to examining the changes in chemical composition of starch reserves during germination. There are a number of reasons for this. If the low-molecular weight products are examined, they consist largely of maltose because of the very rapid combined action of α- and β-amylase on any liberated dextrins. Analyses of the high-molecular weight starch components, on the other hand, show minimal changes because of the ''swamping'' effect that results from a large amount of starch that has not been attacked. Because of this, studies are quite often carried out with model systems consisting of purified enzymes and more clearly defined substrates, allowing a better picture of the individual events which occur in the germinating kernel.

3. Examination of Model Systems

Examination of the breakdown of maize starch by Dunn in 1974 indicated that α- and not β-amylase was able to digest native starch granules.[98] This was confirmed in 1978 by

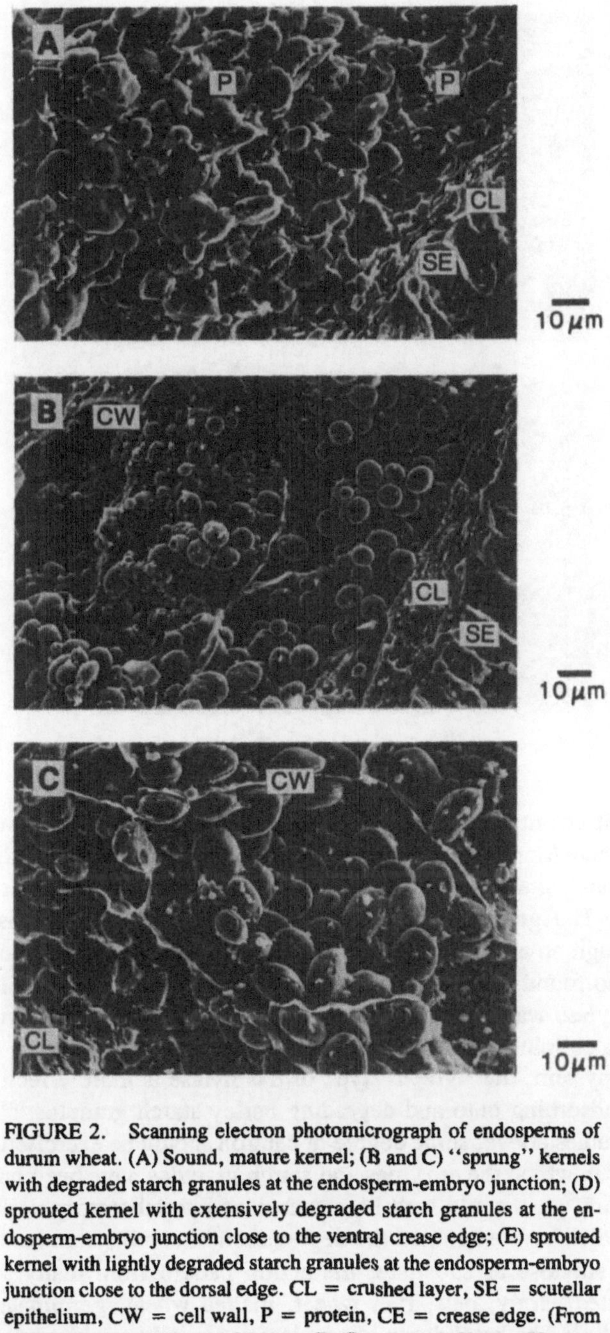

FIGURE 2. Scanning electron photomicrograph of endosperms of durum wheat. (A) Sound, mature kernel; (B and C) "sprung" kernels with degraded starch granules at the endosperm-embryo junction; (D) sprouted kernel with extensively degraded starch granules at the endosperm-embryo junction close to the ventral crease edge; (E) sprouted kernel with lightly degraded starch granules at the endosperm-embryo junction close to the dorsal edge. CL = crushed layer, SE = scutellar epithelium, CW = cell wall, P = protein, CE = crease edge. (From MacGregor, A. W., and Matsuo, R. R., *Cereal Chem.*, 59, 210, 1982. With permission.)

Maeda et al.,[95] who suggested that β-amylase attacked solubilized dextrins liberated by α-amylase digestion.

As far as the relative roles of the "green" or low pI and "germinated" or high pI groups of amylases in this process, varying results have been obtained. Sargeant and Walker in 1978, and Sargeant in 1979, reported that only the germinated or high pI wheat α-amylase was able to adsorb and degrade intact starch granules in wheat.[99,100] The low pI wheat α-amylase presumably assisted by degradation of the solubilized dextrins. Using a purifi-

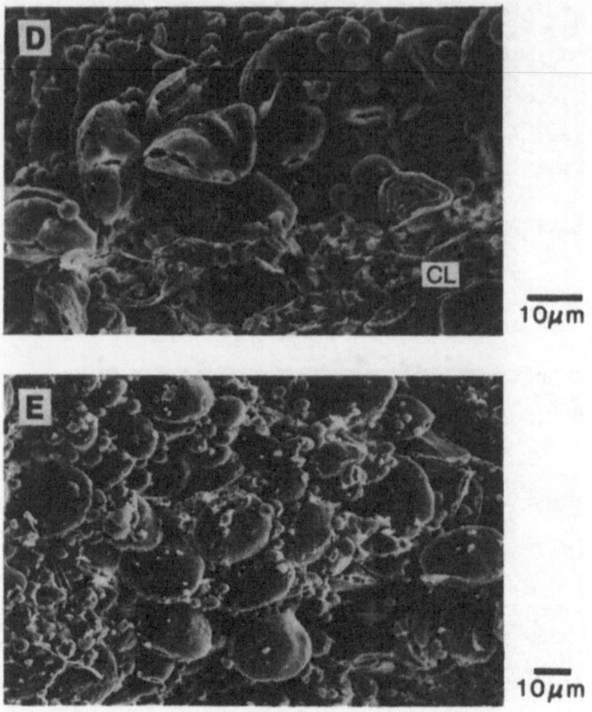

FIGURE 2D and E.

cation consisting of affinity chromatography on cycloheptaamylose followed by chromatofocusing, Kruger and Marchylo in 1985 found that germinated α-amylase was adsorbed preferentially to starch granules at 37°C.[101] Adsorption to small granules was greater than to large granules. The green α-amylases were found, however, to also degrade starch granules, even though no apparent adsorption was found. At low temperatures, the "green" α-amylase was also found to the starch granules. Weselake and Hill in 1983 reported that the "green" α-amylase was adsorbed to starch granules to a greater extent than germinated and effected a greater solubilization of α-amylase.[102]

With the barley system, the "green" type of α-amylase is more effective than the "germinated" type in adsorbing onto and degrading barley starch granules.[103,104]

With all of the studies above, some degree of caution should be exercised, as many factors, such as the manner in which the enzymes and starch granules have been prepared, pH, ionic strength, temperature, etc., could well be a source of interlaboratory variation.

A number of studies have examined the degradation of soluble starch substrates by cereal α-amylases. With amylose as substrate, the action pattern from many sources, including green and germinated barley, oats, rye, wheat, malted wheat, germinated rice, and pearl millet are similar.[105-111] The breakdown of amylose is initially random but becomes increasingly nonrandom with decrease in size of substrate.[109] At least nine subsites could be present at the enzyme active center.[112] As might be expected, there is a slower degradation of amylopectin because of the branched nature of this substrate.[105,111]

Gel permeation experiments have been used to examine the molecular weight changes which result from degradation of soluble starch substrates by α-amylase. Degradation of a β-limit dextrin substrate by hard red spring and durum wheat α-amylases was very similar but differed, as might be expected, from that of a fungal α-amylase.[113] Comparison of the molecular weight breakdown products of β-limit dextrin, amylopectin, and amylase hydrolysis by the "green" and "germinated" wheat α-amylase groups indicated that the two

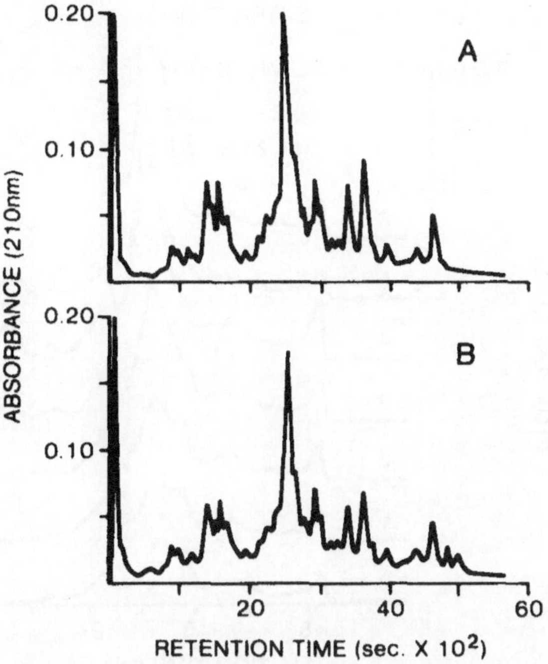

FIGURE 3. RP-HPLC separation of proteins from (A) sound, (B) 96 hr-germinated 'Glenlea' wheat.

groups had very similar action patterns.[101] This suggests very similar functional roles for the two groups.

B. Protein Reserves

1. Microscopic Studies

Typical results using SEM for analyses of changes in protein components are as shown in Figure 2B. It can be observed that breakdown of protein appears to start adjacent to the embryo. Starch granules in this region are essentially free of their protein matrix. With increasing germination, there is a progressive breakdown of this protein matrix.

2. Extraction and Analyses of Protein Components at Different Stages of Germination

At the very early stages of germination, very little breakdown of the storage proteins are observable.[114] Free amino acids are rapidly formed, and this is not surprising since high concentrations of carboxypeptidase exist in the mature endosperm adjacent to the scutellum. Their role is presumably to provide an early source of amino acids for synthesis in the embryo. A recent report indicates that a protease also preexists in the aleurone layer and breaks down aleurone proteins for the same purpose.[115] It would be of interest to determine if there is some regulatory control of germination by this system.

As germination proceeds, there is a gradual breakdown of the storage reserves synchronous with the formation of endoproteolytic activity.[114] A number of observations can be made about this breakdown:

1. The storage proteins degrade simultaneously and very little evidence of large changes in the composition of such proteins is evident.[116] For example, Figure 3 shows the reversed-phase HPLC (RP-HPLC) chromatographic profiles of the gluten proteins of ungerminated and 96-hr germinated 'Glenlea' wheat. It can be seen that the chromatographic profiles are very similar. Recent quantitative analyses of HPLC chro-

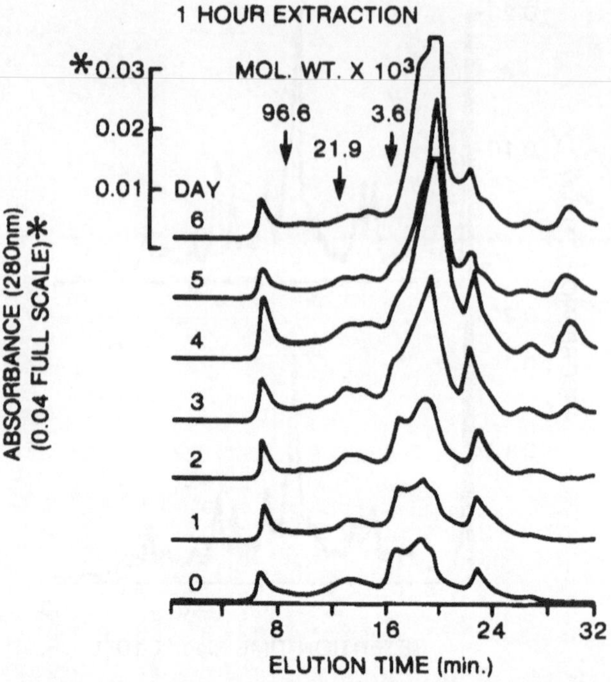

FIGURE 4. Molecular weight distribution of buffer-soluble proteins in 'Glenlea' wheat germinated for up to 6 days. Analyses by HPGPC using a 30 × 0.75 cm Sperogel-TSK 3000 column. Injection volumes were 100 μℓ for wheat germinated up to 4 days and 50 μℓ for wheat germinated for 5 and 6 days.

matograms, however, have detected some preferential breakdown of high-molecular weight glutenins.[15a]

2. The main products which form are low-molecular weight peptides and amino acids, with very little evidence of intermediate molecular weight material.[114,117] For example, changes in the buffer-soluble proteins of 'Glenlea' wheat for a 6-day germination period using high performance gel permeation chromatography (HPGPC) are shown in Figure 4. The major observable changes are in the formation of proteinaceous material of molecular weight less than 3600. Similarly, analysis of changes in the total extractable protein material present in 'Glenlea' wheat indicates that the major changes were found in low-molecular weight material.[116]
3. The increases in amino nitrogen were almost all due to amino acids, with peptides remaining at a fixed level.[114]

There are a few likely explanations to the above phenomena. First of all, the digestion of the endosperm contents proceeds slowly from the embryo end of the kernel. Until sprouting has been extensive, there would be a preponderance of undigested reserves. The large amount of undigested protein reserves could well mask the changes occurring within the sink comprising the digested reserves. Furthermore, the area of initial storage protein mobilization could well be limited to the interface between the digested and undigested reserves. Once limited hydrolysis had occurred due to endoproteolysis, sufficient quantities of exoproteolytic enzymes are present to rapidly hydrolyze the peptide fragments to amino acids.[114] The endo- and exoproteolytic digestion of storage proteins may in fact be similar to the α- and β-amylase hydrolysis of starch where initial scission of the high molecular weight starch components by the α-amylase is followed by rapid breakdown by β-amylase of the dextrin fragments to maltose.

IV. EFFECTS OF PREHARVEST SPROUTING ON END-PRODUCT QUALITY

A number of studies have focused on the chemistry of how quality is affected by preharvest sprouting. The main foodstuff evaluated in this way has been bread. The advantages of such studies are that they provide a real-life situation in which to evaluate the effects of preharvest sprouting and provide the opportunity to try remedial solutions to alleviate deleterious effects. The disadvantage of such studies is that if laboratory-germinated material is used, the situation is somewhat artificial. On the other hand, if field-sprouted samples are used, there may be other factors present in the sample, such as frost, which render unambiguous interpretation difficult.

In all such studies, there is also the uncertainty of what may be causing the observed change in quality. In other words, you cannot separate the relative contribution of enzymes and actual degradation of the endosperm-stored reserves responsible for quality.

Generally, studies may be carried out in two ways:

1. The influence of sprouting may be assessed by tests which relate to end-product quality. For example, with wheat flour, rheological tests are indicative of its absorption and mixing characteristics.
2. The end-product may be prepared and the deterioration in quality subjectively assessed.

A. Bread

In addition to the well-known decrease in falling number or amylograph values of wheats with increasing sprout-damage, a number of other changes due to sprouting can be detected using routine cereal chemistry instruments.[118-124]

On the farinograph, there is a decrease in absorption, a reduction in dough development time, and increase in mechanical tolerance index with increasing sprouting. Sedimentation test values also decrease. With the mixograph, there is a reduction in the mix time needed to obtain optimum dough development. All such tests point, therefore, to weakening of dough properties due to sprouting which should result in a deterioration of bread quality unless the flour is extremely strong. The above effects are largely attributable to changes in the gluten proteins and would suggest that the alterations have been caused by proteolytic enzymes, although other enzymes such as disulfide reductases could be operative. Gluten has been isolated from sprouted wheat and has been found to be extremely sticky.[119,121] Enzymes associated with the gluten have been suggested as causing this effect. In most of such studies, it should be reiterated that the level of sprouting was far beyond what would be considered acceptable levels for use in breadmaking.

It is difficult to pinpoint precise levels of sprouting that can be tolerated in the breadmaking process itself because of the very many types of formulations and breads that are made. For example, the effect of excess α-amylase will be much greater in a long fermentation process than in a short one. There are a number of observations, however, which apply to most types of western style pan breads based on a number of studies.

First of all, some α-amylase is needed to sustain the production of sugars required or proper fermentation and consequent gas production. Millers would prefer, however, to have a low but consistent level of enzymes in their flours and to add a malt or fungal source. Second, excess α-amylase may or may not show up prior to the baking stage of breadmaking. With long fermentation processes, doughs become stickier and harder to handle with increased sprouting. This is presumably due to extensive enzymic hydrolysis of the damaged starch. Such damaged starch has a large water-binding capacity, and its breakdown leads to release of this water which is in excess of that required by the other biochemical constituents, particularly protein, undamaged starch, and pentosans, in the dough.[121,122,125] In the Chorleywood process, however, high energy mixing, rather than fermentation, is used for dough development. In this case, the first real evidence of a problem may show up when the bread is sliced and the well-known stickiness attributable to sprouted wheat shows up in the crumb.[126-128] Other changes result in a loaf having a very high crust color and an open structure. By chemical and physical analyses of bread loaves, Chamberlain and co-workers have carried out extensive studies on chemical changes effected by increased α-amylase in the breadmaking process.[127,128] In general, it was observed that there was an increase in loaf volume and a corresponding decrease in density of a loaf of bread. In the loaf itself, however, there was a density gradient with the inner part decreasing and the outer layers increasing as a result of the increased α-amylase. This density distribution was attributable to an outward movement of solids. As a consequence, poor sliceability resulted because of a loss of crumb mechanical strength. Chemical analyses indicated that the amount of an "amylose" fraction, as well as dextrins, rose with increasing α-amylase. The amylose fraction was greater just under the crust, whereas high-molecular weight dextrins were greatest at the "break" position of these loaves. High concentrations of these components were believed responsible for the stickiness causing slicing problems. Analysis of gums on the blades of the slicer, in fact, indicated that both the "amylose" and high-molecular weight dextrin fraction were present in high concentrations. The explanation proposed for the above behavior was that thermal inactivation of α-amylase occurs rapidly at the crust where temperature is greatest. As a consequence, reduction in molecular size of starch components would be less. High-molecular weight dextrins, on the other hand, would be in greatest concentration at the "break" position of the loaf since crumb at this position comes from the moist interior of the loaf which is at a lower temperature, allowing some degradation of starch granules.

A number of remedial suggestions have been suggested to alleviate the problem. These include cutting back on the α-amylase used as a supplement; decreasing the water absorption; reducing the level of yeast; adding chemicals such as sodium pyrophosphate, acid calcium phosphate, sodium stearyl lactylate; adding extra fat; and insuring that the slicer blades are well lubricated with oil.

The above study describes research using a Chorleywood process. With a hearth bread, such as the type consumed in France, there are unique problems and presumably a different chemistry involved when excess α-amylase is present.[129]

The effects of field-sprouted soft white wheat were evaluated for nine hearth breads.[130] These included Egyptian "balady" bread, Moroccan sour-dough bread, Moroccan straight dough whole wheat flour bread, Pakistani chapatti, Iranian barbari bread, and four types of Indian breads. With the exception of the Iranian barbari bread and the Egyptian "balady" bread, the products seemed very tolerant to sprout damage. Remedial measures such as decreasing water absorption or adding dusting could alleviate processing problems. Whether an unsuspecting user of such products would always be able to foresee such processing problems and be willing to take such preventative measures is another matter. It is possible that the problem with such wheats, in actual practice, may be more serious than suggested.

B. Japanese-Type Sponge Cake

Finney and co-workers have evaluated the effect of α-amylase on Japanese-type sponge cake.[131] Volume initially increased with increasing α-amylase but then decreased. At least 2.5% sprout could be tolerated before quality deteriorated. The crumb and other cake flour quality parameters did not appear to suffer as a result of elevated enzyme levels.

C. Pasta Quality

Japanese noodles appear very sensitive to low levels of sprout damage.[132,133] The observable effects on quality are a wet sticky dough and noodles which tend to droop when hung to dry. Determination of the causative agents for such behavior is enigmatic. Thus, preferential inactivation experiments indicated that protease, α-amylase and, as well, a heat-stable, water-soluble fraction from malted wheat flour all contributed to the effect. Surprisingly, an isolated wheat α-amylase fraction had no effect. More research will be needed in the future to clarify the nature of heat-stable components which could be contributing to the deleterious quality and to establish the relative role of the different enzymes.

Spaghetti quality is also affected by sprouting, but the effects are more subtle, with higher levels of sprout damage being tolerated than in wheat. One reason for this is the limited mobility of water in a pasta product which would inhibit interaction of enzymes such as α-amylase with their substrate.

Studies by Maier in 1980 indicated that use of sprouted durum resulted in a dough with loss of elasticity.[134] The resulting pasta broke apart in the drying process and lacked firmness. Using laboratory-germinated wheat, Kruger and Matsuo in 1982 also found there were increases in the amount of cooking water residue with increase in amount of germination.[135] Increases in reducing sugars were also found in the spaghetti, thereby implicating carbohydrases such as α-amylase. This study was confirmed subsequently, using field-sprouted wheat, by Matsuo et al. in 1982.[136] Sprouted durum increased the residue in the cooking water and increased reducing sugars in both semolina and spaghetti. The spaghetti appeared to be softer when cooked. No changes in the proteins were detected using either the Osborne fractionation scheme or gel filtration, suggesting that proteases were not implicated.

D. Beer

Preharvest sprouting of barley affects the malting and brewing process in a number of ways. According to Brookes,[137] pregermination of barley will have a serious impact on the malting quality of a barley. Shrunken kernels may be lost in the overflow of the steeping tanks. Mold growth may increase, and loss of embryo viability will result in a poorly converted malt and a subsequent low extractability of fermentable material. Furthermore, the extractable material will be of a greater size and more difficult to ferment.

V. CONCLUSIONS

There have been tremendous advances in the biochemistry of sprouting over the last decade. We have a much fuller understanding of the components believed responsible for affecting the quality of end-products. In particular, a number of enzymes such as α-amylase have been rigorously characterized. Much work has been carried out on the multiplicity of such enzymes, and their anatomical location in the germinating kernel is well established. A substantial amount of research has gone into how the quality-important storage reserves of the endosperm have been altered in the germinating process. Although disliking high α-amylase in the breadmaking processes, we have learned somewhat to live with it by modifying the breadmaking process or adding chemicals which partially inhibit the effect of the enzyme.

In spite of the above, we must continue to unravel all the elements present in the biochemical processes of germinating cereals which lead to deterioration of quality. For example, there are other enzymes such as disulfide reductases, glucanases, etc. of which we know very little. We do not fully understand the relative role of enzymes vs. deterioration of reserves in affecting quality. More attention should be directed to understanding the biochemical changes that are occurring in specific end-product processes rather than emphasizing the biochemistry of germination, per se. The area remains a fertile one for research; the next decade should show continued progress and hopefully provide a practical answer to this serious problem.

REFERENCES

1. **Khan, A. A.,** *The Physiology and Biochemistry of Seed Development, Dormancy and Germination*, Elsevier Press, Amsterdam, 1962.
2. **Mayer, A. M. and Poljakoff-Mayber, A.,** *The Germination of Seeds*, 3rd ed., Pergamon Press, Oxford, 1982.
3. **Nozzolillo, C., Lea, P. J., and Loewus, F.,** Mobilization of reserves in germination, in *Recent Advances in Phytochemistry*, Vol. 17, Plenum Press, New York, 1983.
4. **Bewley, J. D. and Block, M.,** *Seeds: Physiology of development and germination*, Plenum Press, New York, 1985.
5. **Hagberg, S.,** Note on a simplified rapid method for determining alpha-amylase activity, *Cereal Chem.*, 38, 202, 1961.
6. **Mottonen, K.,** Comparison of the agar-plate diffusion method for α-amylase assay with the colorimetric stationary-stage method, *J. Sci. Food Agric.*, 21, 261, 1970.
7. **Hejgaard, J. and Gibbons, G. C.,** Screening for α-amylase in cereals. Improved gel-diffusion assay using a dye-labelled starch substrate, *Carlsberg Res. Commun.*, 44, 21, 1979.
8. **Barnes, W. C. and Blakeney, A. B.,** Determination of cereal α-amylase using a commercially available dye-labelled substrate, *Starch/Staerke*, 26, 193, 1974.
9. **Barnes, W. C.,** New quality tester detects rain-damaged wheat at silos, *Agric. Gaz. N.S.W.*, 90, No. 5, 1979.
10. **Mathewson, P. R. and Pomeranz, Y.,** Detection of sprouted wheat by a rapid colorimetric determination of α-amylase, *J. Assoc. Off. Agric. Chem.*, 60, 16, 1977.
11. **Tipples, K. H.,** A viscometric method for measuring alpha-amylase activity in small samples of wheat and flour, *Cereal Chem.*, 46, 589, 1969.
12. **Kruger, J. E. and Tipples, K. H.,** A modified procedure for use of the Perkin Elmer Model 191 Grain Amylase Analyzer in determining low levels of α-amylase in wheats and flours, *Cereal Chem.*, 58, 271, 1981.
13. **Sandstedt, R. M., Kneen, E., and Blish, M. J.,** A standardized Wohlgemuth procedure for alpha-amylase activity, *Cereal Chem.*, 16, 1712, 1939.
14. **Kruger, J. E. and Marchylo, B. A.,** The use of reduced β-limit dextrin as substrate in an automated amylase assay, *Cereal Chem.*, 49, 453, 1972.
15. **Kruger, J. E.,** Changes in the amylases of hard red spring wheat during growth and maturation, *Cereal Chem.*, 49, 379, 1972.

15a. **Kruger, J. E.,** unpublished.

16. **Kruger, J. E. and Tipples, K. H.,** Relationships between falling number, amylograph viscosity and α-amylase activity in Canadian wheat, *Cereal Res. Commun.*, 8, 97, 1980.
17. **Kruger, J. E.,** Severity of sprouting as a factor influencing the distribution of α-amylase in pilot mill streams, *Can. J. Plant Sci.*, 61, 817, 1981.
18. **Henry, R. J., Martin, D. J., and Blakeney, A. B.,** Pearling and milling as techniques for the improvement of the quality of sprouted wheat, in *Proc. 4th Int. Symp. Pre-harvest Sprouting*, D. Mares, Ed., Westview Press, Boulder, Colo., 1987, 207.
19. **Lui, R., Liang, Z., Posner, E. S., and Ponte, J. G., Jr.,** A technique to improve functionality of flour from sprouted wheat, *Cereal Chem.*, 7, 471, 1986.
20. **Alexandrescu, V. and Mihailescu, F.,** Electrophoretic study of wheat α-amylases during the first growing stages, *Rev. Roum. Biochim.*, 7, 3, 1970.
21. **Olered, R. and Jonsson, G.,** Electrophoretic studies of alpha-amylases in wheat, *J. Sci. Food Agric.*, 21, 385, 1970.
22. **Marchylo, B., LaCroix, L. J., and Kruger, J. E.,** α-Amylase isoenzymes in Canadian wheat cultivars during kernel growth and maturation, *Can. J. Plant Sci.*, 60, 433, 1980.
23. **Weselake, R. J., MacGregor, A. W., and Hill, R. D.,** An endogenous α-amylase inhibitor in barley kernels, *Plant Physiol.*, 72, 809, 1983.
24. **Weselake, R. J., MacGregor, A. W., Hill, R. D., and Duckworth, H. W.,** Purification and characteristics of an endogenous α-amylase inhibitor from barley kernels, *Plant Physiol.*, 73, 1008, 1984.
25. **Marchylo, B. A. and Kruger, J. E.,** Degradation of starch granules in maturing wheat and its relationship to alpha-amylase production by the embryo, in *Proc. 4th Int. Symp. Pre-Harvest Sprouting*, D. Mares, Ed., Westview Press, 1987, 483.
26. **MacGregor, A. W., Thompson, R. G., and Meredith, W. O. S.,** α-Amylase from immature barley: purification and properties, *J. Inst. Brew. London*, 80, 181, 1974.
27. **Marchylo, B. A., Kruger, J. E., and Irvine, G. N.,** α-Amylase from immature hard red spring wheat. I. Purification and some chemical and physical properties, *Cereal Chem.*, 53, 157, 1976.
28. **Tkachuk, R. and Kruger, J. E.,** Wheat α-amylase. II. Physical characterization, *Cereal Chem.*, 51, 508, 1974.

29. **Bernfeld, O.,** Amylases, α and β, in *Methods in Enzymology,* Colowick, S. O., and Kaplan, N. O., Eds., Vol. 1, Academic Press, New York, 1955, 149.
30. **Bureau, D. and MacGregor, A. W.,** Evaluation of the neocuproine method for determining reducing sugars. Application for the determination of the activity of barley beta-amylase, *Lebensm. Wiss.-Technol.,* 13, 92, 1980.
31. **Warchalewski, J. R. and Tkachuk, R.,** Durum wheat α-amylases: Isolation and purification, *Cereal Chem.,* 55, 146, 1978.
32. **Mathewson, P. R. and Seabourn, W.,** A new procedure for specific determination of β-amylase in cereals, *J. Agric. Food Chem.,* 31, 1322, 1983.
33. **Rowsell, E. V. and Goad, L. J.,** The constituent of wheat binding latent β-amylase, *Biochem. J.,* 84, 73P, 1962a.
34. **Rowsell, E. V. and Goad, L. J.,** Latent β-amylase of wheat: Its mode of attachment to glutenin and its release, *Biochem. J.,* 84, 74P, 1962.
35. **Okamoto, K. and Akazawa, T.,** Enzymic mechanism of starch breakdown in germinating rice seeds. IX. De novo synthesis of β-amylase, *Plant Physiol.,* 54, 81, 1980.
36. **Tkachuk, R. and Tipples, K. H.,** Wheat β-amylase. II. Characterization, *Cereal Chem.,* 43, 62, 1966.
37. **Kruger, J. E.,** On the relationship between free and latent β-amylases in wheat, *Cereal Chem.,* 47, 49, 1970.
38. **Kruger, J. E.,** Modification of wheat β-amylase by proteolytic enzymes, *Cereal Chem.,* 56, 298, 1979.
39. **Hejgaard, R. J.,** "Free" and "bound" β-amylases during malting of barley. Characterization by two-dimensional immunoelectrophoresis, *J. Inst. Brew. London,* 84, 43, 1978.
40. **LaBerge, D. E. and Marchylo, B. A.,** Changes in beta-amylase enzymes during kernel development of barley and the effect of papain as an extractant, *Am. Soc. Brew. Chem. Proc.,* 44, 16, 1986.
41. **Meyer, K. H., Spahr, P. F., and Fischer, E. H.,** Purification, crystallisation et propriétés de la beta-amylase de blé. Sur les enzymes amylolytiques 22, *Helv. Chim. Acta.,* 36, 1924, 1953.
42. **Danielsson, C. E.,** Molecular weight of malt-amylase, *Nature (London),* 162, 525, 1948.
43. **Maeda, I., Jimi, N., Taniguchi, H., and Nakumura, M.,** Purification of R-enzyme from malted barley and its role in in vitro digestion of barley starch granules, *J. Jpn. Soc. Starch Sci.,* 26, 117, 1979.
44. **Manners, D. J. and Yellowlees, D.,** Studies on debranching enzymes. I. The limit of dextrinase activity of extracts of certain higher plants and commercial malts, *J. Inst. Brew. London,* 79, 377, 1973.
45. **Lee, W. J. and Pyler, R. E.,** Barley malt limit dextrinase: Varietal, environmental and malting effects, *Am. Soc. Brewing Chem.,* 42, 11, 1984.
46. **Hardie, D. G. and Manners, D. J.,** A viscometric assay for pullanase-type, debranching enzymes, *Carbohydr. Res.,* 36, 207, 1974.
47. **Catley, B. J.,** An assay for pullulanase in the presence of other carbohydrases, *Carbohydr. Res.,* 61, 419, 1978.
48. **Kruger, J. E. and Marchylo, B. A.,** Notes on the presence of debranching enzymes in immature wheat kernels, *Cereal Chem.,* 55, 529, 1978.
49. **Yamada, J.,** Purification of oat debranching enzyme and occurrence of inactive debranching enzyme in cereals, *Agric. Biol. Chem.,* 45, 1013, 1981.
50. **Hardie, D. G.,** Control of carbohydrase formation by gibberellic acid in barley endosperm, *Phytochemistry,* 14, 1719, 1975.
51. **Lenoir, P., MacGregor, A. W., Moll, M., and Daussant, J.,** Identification de la R-enzyme de l'orge et du malt par fucalisation isoélectrique, *C. R. Acad. Sci. Paris, t.* 298, Ser. III, 8, 243, 1984.
52. **Hardie, D. G., Manners, D. J., and Yellowlees, D.,** The limit dextrinase from malted sorghum (sorghum vulgare), *Carbohydr. Res.,* 50, 75, 1976.
53. **Iwaki, K., and Fuwa, H.,** Purification and some properties of debranching enzyme of germinating rice endosperm, *Agric. Biol. Chem.,* 45, 2683, 1981.
54. **Sopanen, T., Burston, D., and Matthews, D. M.,** Uptake of small peptides by the scutellum of germinating barley, *FEBS Lett.,* 79, 4, 1977.
55. **Mikola, J. and Kohlemainen, L.,** Localization and activity of various peptidases in germinating barley, *Planta,* 104, 167, 1972.
56. **Kruger, J. E. and Preston, K.,** The distribution of carboxypeptidases in anatomical tissues of developing and germinating wheat kernels, *Cereal Chem.,* 54, 167, 1977.
57. **Preston, K. R. and Kruger, J. E.,** Purification and properties of two proteolytic enzymes with carboxypeptides activity in germinated wheat, *Plant Physiol.,* 58, 516, 1976.
58. **Preston, K. R. and Kruger, J. E.,** Specificity of two isolated wheat carboxypeptidases, *Phytochemistry,* 16, 525, 1977.
59. **Kruger, J. E.,** Effects of proteolytic enzymes on gluten as measured by a stretching test, *Cereal Chem.,* 48, 121, 1971.
60. **Finley, J. W.,** Sulfanilamide-azogluten: A chromophoric gluten derivative for colorimetrically estimating protease activity on gluten, *Cereal Chem.,* 52, 714, 1975.

61. **Kruger, J. E.**, Changes in the levels of proteolytic enzymes from hard red spring wheat during growth and maturation, *Cereal Chem.*, 50, 122, 1973.
62. **Kruger, J. E. and Preston, K.**, The nature and role of proteolytic enzymes during early germination, Proc. 1st Int. Symposium on Pre-harvest Sprouting in Cereals, *Cereal Res. Commun.*, 4, 213, 1976.
63. **Jacobsen, J. V. and Varner, J. E.**, Gibberellic acid-induced synthesis of protease by isolated aleurone layers of barley, *Plant Physiol.*, 42, 1596, 1967.
64. **Preston, K. R. and Kruger, J. E.**, Physiological control of exo- and endproteolytic activities in terminating wheat and their relationship to storage protein hydrolysis, *Plant Physiol.*, 64, 450, 1979.
65. **Sundblom, N. O. and Mikola, J.**, On the nature of the proteinases secreted by the aleurone layer of barley grain, *Physiol. Plant.*, 27, 281, 1972.
66. **Burger, W. C. and Schroeder, R. L.**, A sensitive method for detecting endopeptidases in electrofocused thin-layer gels, *Anal. Biochem.*, 71, 384, 1976.
67. **Wang, C. C. and Grant, D. R.**, The proteolytic enzymes of wheat flour, *Cereal Chem.*, 46, 537, 1969.
68. **Skupin, J. and Warchalewski, J.**, Isolation and properties of protease A from wheat grain, *J. Sci. Food Agric.*, 22, 11, 1971.
69. **Kaminski, E. and Bushuk, W.**, Wheat proteases. I. Separation and detection by starch-gel electrophoresis, *Cereal Chem.*, 46, 317, 1969.
70. **McDonald, C. E. and Chen, L. L.**, Properties of wheat flour proteinases, *Cereal Chem.*, 41, 443, 1964.
71. **Mountfield, J. D.**, LXXX. The proteolytic enzymes of sprouted wheat. I., *Biochem. J.*, 30, 549, 1936.
72. **Mountfield, J. D.**, CCXLIX. The proteolytic enzymes of sprouted wheat. II., *Biochem. J.*, 30, 1778, 1936.
73. **Visuri, K., Mikola, J., and Enari, T. M.**, Isolation and partial characterization of a carboxypeptidase from barley, *Eur. J. Biochem.*, 7, 193, 1969.
74. **Preston, K.**, An automated fluorometric assay for proteolytic activity in cereals, *Cereal Chem.*, 52, 451, 1975.
75. **Preston, K. and Kruger, J.**, Location and activity of proteolytic enzymes in developing wheat kernels, *Can. J. Plant Sci.*, 56, 217, 1976.
76. **Winspear, M. J., Preston, K. R., Rostogi, V., and Oaks, A.**, Comparison of peptide hydrolase activities in cereals, *Plant Physiol.*, 75, 480, 1984.
77. **Ray, L. E.**, Large scale isolation and partial characterization of some carboxypeptidases from malted barley, *Carlsberg Res. Commun.*, 41, 169, 1976.
78. **Baxter, E. D.**, Purification and properties of malt carboxypeptidases attacking hordein, *J. Inst. Brew. London*, 84, 271, 1978.
79. **Mikola, L. and Mikola, J.**, Mobilization of proline in the starchy endosperm of germinating barley grain, *Planta*, 149, 149, 1980.
80. **Kruger, J. E.**, Purification and some properties of malted wheat BAPA-ase, *Cereal Chem.*, 48, 512, 1971.
81. **Sullivan, B. and Howe, M. A.**, Lipases of wheat. I., *J. Am. Chem. Soc.*, 55, 320, 1933.
82. **Drapron, R., Anh, N. X., Launay, B., and Guilbot, A.**, Development and distribution of wheat lipase activity during the course of germination, *Cereal Chem.*, 46, 647, 1969.
83. **Tavener, R. J. A. and Laidman, D. L.**, The induction of lipase activity in the germinating wheat grain, *Phytochemistry*, 11, 989, 1972.
84. **Taneja, S. R., Abrol, Y. P., and Sachar, R. C.**, Modulation of *o*-diphenolase and monophenolase enzymes during wheat grain development, *Cereal Chem.*, 51, 457, 1974.
85. **Taneja, S. R. and Sachar, R. C.**, Induction of polyphenol oxidase in germinating wheat seeds, *Phytochemistry*, 13, 2695, 1974.
86. **Taneja, S. R. and Sachar, R. C.**, Separate monophenolase and *o*-diphenolase enzymes in *Triticum aestivum*, *Phytochemistry*, 13, 1367, 1976.
87. **Kruger, J. E.**, Changes in the polyphenoloxidases of wheat during kernel growth and maturation, *Cereal Chem.*, 53, 201, 1976.
88. **Kruger, J. E. and LaBerge, D. E.**, Changes in peroxidase activity and peroxidase isoenzymes of wheat during germination, *Cereal Chem.*, 51, 578, 1974.
89. **Kruger, J. E.**, Changes in the catalases of wheat during kernel growth and maturation, *Cereal Chem.*, 54, 820, 1977.
90. **Evers, A. D. and McDermott, E. E.**, Scanning electron microscopy of wheat starch. II. Structure of granules modified by α-amylolysis — Preliminary report, *Staerke*, 22, 23, 1970.
91. **Dronzek, B. L., Hwang, P., and Bushuk, W.**, Scanning electron microscopy of starch from sprouted wheat, *Cereal Chem.*, 49, 232, 1972.
92. **Jones, F. T. and Bean, M. M.**, A light and SEM look at enzyme-damaged wheat starch, *Microscope*, 20, 33, 1972.
93. **Palmer, G. H.**, Morphology of starch granules in cereal grains and malts, *J. Inst. Brew. London*, 78, 326, 1972.

94. **Lineback, D. R. and Ponpipom, S.**, Effects of germination of wheat, oats and pearl millet on α-amylase activity and starch degradation, *Staerke*, 29, 52, 1977.
95. **Maeda, I., Kiribuchi, S., and Nakamura, M.**, Digestion of barley starch granules by the combined action of α- and β-amylases purified from barley and barley malt, *Agric. Biol. Chem.*, 42, 259, 1978.
96. **MacGregor, A. W. and Matsuo, R. R.**, Starch degradation in endosperms of barley and wheat kernels during initial stages of germination, *Cereal Chem.*, 59, 210, 1982.
97. **Bathgate, G. N. and Palmer, G. H.**, The in vivo and in vitro degradation of barley and malt starch granules, *J. Inst. Brew. London*, 79, 402, 1973.
98. **Dunn, G.**, A model for starch breakdown in higher plants, *Phytochemistry*, 13, 1341, 1974.
99. **Sargeant, J. G. and Walker, T. S.**, Adsorption of wheat α-amylase isoenzymes to wheat starch, *Staerke*, 30, 160, 1978.
100. **Sargeant, J. G.**, The α-amylase isoenzymes of developing and germinating wheat grain, in *Recent Advances in the Biochemistry of Cereals*, Laidman, D. L., and Wyn-Jones, R. G., Eds., Academic Press, London, 1979, 339.
101. **Kruger, J. E. and Marchylo, B. A.**, A comparison of the catalysis of starch components by isoenzyme from the two major groups of germinated wheat α-amylase, *Cereal Chem.*, 62, 11, 1985.
102. **Weselake, R. J. and Hill, R. D.**, Inhibition of alpha-amylase catalyzed starch granule hydrolysis by cyclohepaamylose, *Cereal Chem.*, 60, 98, 1983.
103. **MacGregor, A. W. and Ballance, D. L.**, Quantitative determination of α-amylase separation by isoelectric focusing, *J. Inst. Brew. London*, 86, 131, 1980.
104. **Bertoft, E. and Kulp, S.-E.**, A gel filtration study on the action of barley α-amylase isoenzymes on granular starch, *J. Inst. Brew. London*, 92, 64, 1986.
105. **Greenwood, C. T. and MacGregor, A. W.**, The isolation of α-amylase from barley and malted barley, and a study of the properties and action-patterns of the enzymes, *J. Inst. Brew. London*, 71, 405, 1965.
106. **Greenwood, C. T. and Milne, E. A.**, Studies in starch degrading enzymes. VII. Properties and action-pattern of the amylases from barley, oats, rye 1 and wheat, *Staerke*, 20, 101, 1968.
107. **Greenwood, C. T. and Milne, E. A.**, Studies on starch-degrading enzymes. VIII. A comparison of α-amylases from different sources, their properties and action patterns, *Staerke*, 5, 139, 1968.
108. **MacGregor, A. W.**, α-Amylase I from malted barley — physical properties and action pattern on amylose, *Cereal Chem.*, 5, 754, 1978.
109. **MacGregor, A. W.**, Cereal α-Amylases: synthesis and action pattern, in *Seed Proteins*, Daussant, J., Mossé, J., and Vaughan, J., Eds., Academic Press, London, 1983.
110. **Murata, T., Akazawa, F., and Fukuchi, S.**, Enzymic mechanism of starch breakdown in germinating rice seeds. I. An analytical study, *Plant Physiol.*, 43, 1899, 1968.
111. **Beleia, A. and Varriano-Marston, E.**, Pearl millet amylases. I. Properties of partially purified alpha-amylase, *Cereal Chem.*, 58, 433, 1981.
112. **MacGregor, E. A. and MacGregor, A. W.**, The action of cereal α-amylases on solubilized starch and cereal starch granules, in *New Approaches to Research on Cereal Carbohydrates*, Hill, R. D. and Munck, L., Eds., Elsevier, Amsterdam, 1983, 149.
113. **Kruger, J. E. and Marchylo, B. A.**, High performance aqueous gel permeation chromatographic analysis of β-limit dextrin hydrolysis by malted hard red spring wheat, malted durum wheat and fungal (*Aspergillus oryzae*) α-amylases, *Cereal Chem.*, 59, 488, 1982.
114. **Preston, K. R., Dexter, J. E., and Kruger, J. E.**, Relationship of exo- and endoproteolytic activity to storage protein hydrolysis in germinating durum and hard spring wheat, *Cereal Chem.*, 55, 877, 1978.
115. **Jofre-Garfias, A., Garcia-Maya, M., and Hamabata, A.**, On the amino acid source for α-amylase synthesis induced by GA_3 in wheat aleurone layers, *Plant Physiol.*, 75(1), abstr. 1052, 1984.
116. **Kruger, J. E. and Marchylo, B. A.**, Examination of the mobilization of storage proteins of wheat kernels during germination by high-performance reversed-phase gel permeation chromatography, *Cereal Chem.*, 62, 1, 1985.
117. **Kruger, J. E.**, Rapid analysis of changes in the molecular weight distribution of buffer soluble proteins during germination of wheat, *Cereal Chem.*, 61, 205, 1984.
118. **Finney, K. F., Natsuaki, O., Bolte, L. C., Mathewson, P. R., and Pomeranz, Y.**, Alpha-amylase in field-sprouted wheats: its distribution and effect of Japanese-type sponge cake and related physical and chemical tests, *Cereal Chem.*, 58(4), 355, 1981.
119. **Ciacco, C. F. and D'Appolonia, B. L.**, Reconstitution studies with sound and sprouted wheat flour, *Cereal Chem.*, 59(2), 77, 1982.
120. **D'Appolonia, B. L.**, 'Sprouted' flour coping with damage, *Bakers Digest, Milling Baking News*, 6, 6, 1983.
121. **Kulp, K., Roewe-Smith, P., and Lorenz, K.**, Preharvest sprouting of winter wheat. I. Rheological properties of flours and physicochemical characteristics of starches, *Cereal Chem.*, 60(5), 355, 1983.
122. **Lorenz, K., Roewe-Smith, P., Kulp, K., and Bates, L.**, Preharvest sprouting of winter wheat. II. Amino acid composition and functionality of flour and flour fractions, *Cereal Chem.*, 60(5), 360, 1983.

123. **Lukow, O. M., and Bushuk, W.,** Influence of germination of wheat quality. II. Modification of endosperm protein, *Cereal Chem.*, 61(4), 340, 1984.
124. **Lukow, O. M. and Bushuk, W.,** Influence of germination of wheat quality. I. Functional (breadmaking) and biochemical properties, *Cereal Chem.*, 61(4), 336, 1984.
125. **Ibrahim, Y. and D'Appolonia, B. L.,** Sprouting in hard red spring wheat, *Bakers Digest*, October, 1979, p. 17.
126. **Buchanan, A. M. and Nicholas, E. M.,** Sprouting, alpha-amylase and breadmaking quality, *Cereal Res. Commun.*, 8, 23, 1980.
127. **Chamberlain, N., Collins, T. H., and McDermott, E. E.,** Alpha-amylase and bread properties, *J. Food Technol.*, 16, 127, 1981.
128. **Chamberlain, N., Collins, T. H., and McDermott, E. E.,** The influence of α-amylase on loaf properties in the UK, *Proc. 7th World Cereal and Bread Congress*, Holas, J. and Kratochvil, J., Eds., Elsevier Press, Amsterdam, 1982, 841.
129. **Grandvoinnet, P. and Berger, M.,** Breadmaking of French sprout damaged wheats, *Proc. 7th World and Bread Congress*, Holas, J. and Kratochval, J., Eds., Elsevier Press, Amsterdam, 1982, 837.
130. **Finney, P. L., Morad, M. M., Patel, K., Chaudhry, S. M., Ghiasi, K., Ranhotra, G., Seitz, L. M., and Sebti, S.,** Nine international breads from sound and highly-field-sprouted Pacific Northwest soft white wheat, *Baker's Digest*, 6, 22, 1980.
131. **Finney, K. F., Natsuaki, O., Bolte, L. C., Mathewson, P. R., and Pomeranz, Y.,** Alpha-amylase in field-sprouted wheats: Its distribution and effect on Japanese-type sponge cake and related physical and chemical tests, *Cereal Chem.*, 58, 355, 1981.
132. **Bean, M. M., Keagy, P. M., Fullington, J. G., Jones, F. T., and Mecham, D. K.,** Dried Japanese noodles. I. Properties of laboratory-prepared noodle doughs from sound and damaged wheat flours, *Cereal Chem.*, 51, 417, 1974.
133. **Bean, M. M., Nimmo, C. C., Fullington, J. G., Keagy, P. M., and Mecham, D. K.,** Dried Japanese noodles. II. Effect of amylase, protease, salts, and pH on noodle doughs, *Cereal Chem.*, 51, 427, 1974.
134. **Maier, M. G.,** Widespread sprout damage, *Macaroni J.*, 62(10), 20, 1980.
135. **Kruger, J. E. and Matsuo, R. R.,** Comparison of alpha-amylase and simple sugar levels in sound and germinated durum wheat during pasta processing and spaghetti cooking, *Cereal Chem.*, 59, 26, 1982.
136. **Matsuo, R. R., Dexter, J. E., and MacGregor, A. W.,** Effect of sprout damage on durum wheat and spaghetti quality, *Cereal Chem.*, 59, 468, 1982.
137. **Brookes, P. A.,** The significance of pre-harvest sprouting of barley in malting and brewing, *Cereal Res. Commun.*, 8, 25, 1980.

Chapter 5

THE GENETICS OF PREHARVEST SPROUTING IN CEREALS, PARTICULARLY IN WHEAT

M. D. Gale

TABLE OF CONTENTS

I. INTRODUCTION

The study of preharvest sprouting (PHS) phenomena in cereals, particularly wheat, has a long history, usually motivated by the need to develop breeders' screens for resistant genotypes. Nevertheless, our knowledge of the genetic control of the syndrome is still extremely fragmentary. There are several reasons for our apparently slow progress.

The first is that which has dogged all genetic studies in wheat. Genetically, *Triticum aestivum* ($2n = 6x = 42$) is complex. It is an allohexaploid with three sets, or genomes, of seven pairs of chromosomes derived from at least three diploid progenitors. These are known as the A, B, and D genomes. Although chromosome pairing, and hence recombination, between chromosomes in different genomes, does not usually occur, there is considerable similarity, or homoeology, between the different genomes. This is because each of the three diploid ancestors were themselves originally derived from a single primeval species. Furthermore, because wheat is such a recently evolved species, possibly only 5000 years old,[1] diploidization has not yet occurred to the extent found in more ancient polyploid species. The result is that for any gene in wheat, there are likely to be two other active loci on corresponding chromosomes in the remaining genomes. The best evidence for this derives from our knowledge of biochemical markers where the gene lists show that in many cases functional genes coding for homologous proteins are to be found in each of the three wheat genomes.[2] Thus, it is to be expected that genes affecting the preharvest sprouting phenotype will be similarly triplicated. Single gene control which may be simple in diploid Triticeae species, such as rye or barley, can in wheat be so complex that its genetic analysis can only be handled by the techniques of quantitative genetics.

Second, our appreciation of the complexity of the preharvest sprouting character has been slow to evolve. Attempts to interpret analyses of segregating families for parameters which might have value as breeders' screens have often defied simple genetic explanation. For example, many studies have involved measurements of the Hagberg falling number in grain from plants allowed to stand in the field for an extended period following harvest ripeness; it is now clear that differences among such data may result from several sources, including variation in α-amylase accumulation prior to harvest ripeness, variation in the duration of dormancy following harvest ripeness, and variation in the rate that PHS damage is sustained following the decay of dormancy. Still further levels of complexity may derive from genes affecting different physiological systems which may impinge on any single source of variation. For these reasons, the most effective genetic analyses have been for those component characters which are underpinned by a strong base of physiological understanding. In this chapter an attempt has been made to compile a review of genetic analyses of these individual components, but in several cases it has been necessary to point out limitations and confounding sources of genetic variation.

Finally, in defense of our slow progress in this area, two more factors that have probably hindered the advance of our appreciation of genetic control deserve mention. First, preharvest sprouting is probably, of all characters of interest to breeders, that most affected by environment. Temperature, probably at precise windows during grain development and following grain maturity, and the pattern of availability of moisture over an 8- or 10-week period can make the difference between no sprouting damage at harvest or the complete loss of a crop for the quality market. In genetic analyses of field-grown material in any single season, any variation, from whatever cause, in anthesis time, time of harvest ripeness, or time of harvest can lead to large interactions between genotype and the critical components of environmental control. When considered over different seasons, the problem becomes even worse. Apart from growing all experimental material under controlled environmental conditions, little can be done to remove the extraneous experimental noise, except to carefully consider the possible effects of genotype × environment interactions in all field experiments. Second, the preharvest sprouting character is unusually complex in that genes involved in its regulation can be expressed in any one of three distinct tissues: the maternal plant, the endosperm, or the embryo of the grains which, genetically, belong to the next generation.

In this chapter the current appreciation of the many components of the complex character, preharvest sprouting, are reviewed. The value of this information, as it emerges, is to allow breeders, for whom different components will assume different levels of importance, to choose resistance mechanisms and to select those parents which are most relevant and most easily handled in their individual situations.

II. MATERNAL, DIPLOID, AND TRIPLOID CHARACTERS

In an analysis of any PHS component character, the source of a physiological effect may be either from the maternal tissue or from within the grain itself. The precise source may affect the strategy chosen for analysis and its handling in a practical breeding program.

Maternal tissue includes the entire vegetative plant, up to and including the glumes in the wheat ear or the hulls on a barley grain, and the pericarp surrounding the grain. The grain itself comprises two genetically distinct tissues. The endosperm, which includes the aleurone layer and the starchy endosperm, is triploid and has two haploid sets of chromosomes from the maternal parent and one from the pollen parent. The embryo tissues are diploid, and have the same genotype as the endosperm except for the extra maternal set of chromosomes.

Maternal characters, such as red grain color which is present in the testa layer of the pericarp, will not segregate until the F_2 generation is grown, whereas zygotic characters will segregate in the F_2 seed on the F_1 plant. Similarly, reciprocally backcrossed grain will exhibit reciprocal differences for maternal characters. Incidentally, maternal control should not be confused with cytoplasmic inheritance where reciprocal differences will be maintained in the F_2 and beyond. Triploid endosperm characters will segregate in the F_2 grain on an F_1 plant, but in a 1:1:1:1 ratio, which includes two heterozygous classes, *AAa* and *aaA*. Similarly, reciprocal differences, not seen for embryo characters, can occur for endosperm characters in backcrosses. Genes expressed in the embryo will exhibit normal Mendelian inheritance and segregation in the F_2 grain on an F_1 plant.

These different segregations for control by cytoplasmic, maternal, endosperm, or embryo characters are shown in Table 1. It is clear that for a single gene with heritable qualitative effects, observation of the phenotypic classes in these generations will allow the elucidation of the histological source of gene action and statements to be made concerning dominance and/or gene additivity. The analysis by Stoy and Olsen for the grain germination response to catechin tannins provides an example of this approach.[3]

In cases where heritability is not so high and the character may only be measured in terms of generation means, progress can still be made by adapting the methods of quantitative analysis adopted by R. A. Fisher and developed by K. Mather and colleagues.[4,5] A triploid scale of inheritance can be constructed using the conventional "d" (additive) and "h" (dominance) components of variation.[6] The main difference in the endosperm model, shown in Figure 1, is that heterozygotes can deviate from the mid-parent even in the absence of dominance, and that there are two distinct intra-allele interaction components, "h1" and "h2". An example of the use of this approach was provided by an analysis of generation means for GA-induced α-amylase responses in parental and backcross generations, where the fitting of alternative diploid and triploid models demonstrated clearly that the character was endosperm mediated and that the expression was allele-dose dependent, with partial dominance for an insensitive response.[6] Table 2 gives examples of the expected family mean components in the two models.

These methods of analysis have not been widely exploited. However, in those cases where genetic variation can be assessed in first generation-crossed grain, such approaches can aid

Table 1
EXPECTED GENOTYPES FOR GRAIN CHARACTERS CONTROLLED BY CYTOPLASMIC GENES, OR BY NUCLEAR GENES EXPRESSED IN THE MATERNAL PARENT VEGETATIVE TISSUES, THE ENDOSPERM, OR THE EMBRYO

Family	Cytoplasm	Maternal plant	Grain: Endosperm	Grain: Embryo
P_1	C_1	AA	AAA	AA
P_2	C_2	aa	aaa	aa
F_1 (1.2)	C_1	AA	AAa	Aa
F_1 (2.1)	C_2	aa	aaA	
F_2 (1.2)	C_1	Aa	AAA:AAa:Aaa:aaa	AA:2Aa:aa
F_2 (2.1)	C_2			
B1 (1.[1.2])	Cl	AA	AAA:AAa	AA:Aa
B1 (1.[2.1])				
B1 ([1.2].1)	C_1	Aa	AAA:aaA	
B1 ([2.1].1)	C_2			
B2 (2.[1.2])	C_2	aa	aaA:aaa	Aa:aa
B2 (2.[2.1])				
B2 ([1.2].2)	C_1	Aa	AAa:aaa	
B2 ([2.1].2)	C_2			

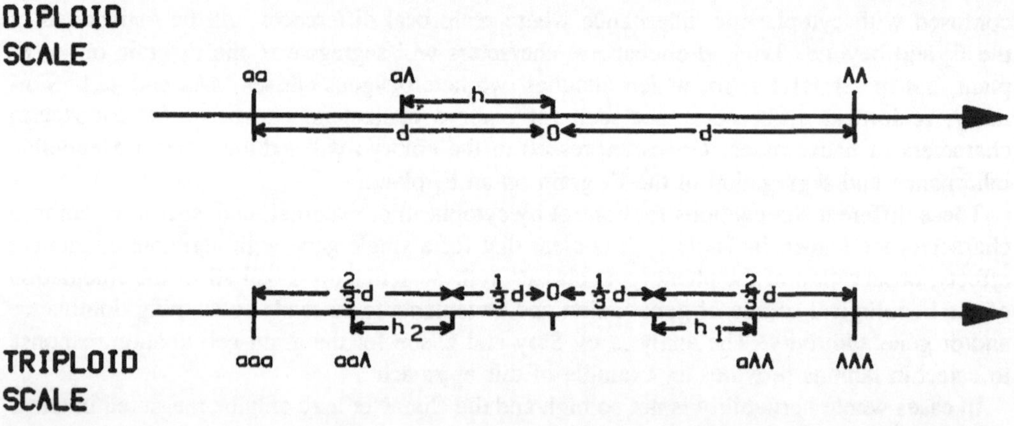

FIGURE 1. Additive (d) and intra-allelic interactive (h) components of variation expected for segregation at a single locus, A/a, in diploid or triploid tissues.

investigations of physiological mechanisms and simplify apparently complex patterns of inheritance.

III. THE PREHARVEST SPROUTING PHENOTYPE

Preharvest sprouting is generally associated with two indicators, the presence of unacceptably high levels of α-amylase and the presence of an unacceptable proportion of visibly sprouted grain in the harvested crop. The range of genetic and environmental sources of variation that have been ascribed to the various physiological components of preharvest sprouting are summarized in Figure 2. This rather uncritical survey of the literature includes

Table 2
FAMILY MEANS USING THE "d" AND "h" COMPONENTS OF VARIANCE DESCRIBED IN FIGURE 1

	Diploid (embryo)	Triploid (endosperm)
P_1	d	d
P_2	−d	−d
F_1(1.2)	h	1/3 d + h1
F_1 (2.1)		−1/3 d + h2
F_2	1/2 h	1/4 h1 + 1/4 h2
B1 (1.[1.2])	1/2 d + 1/2 h	2/3 d + 1/2 h1
B1 ([1.2].1)		1/3 d + 1/2 h2
B2 (2.[1.2])	−1/2 d + 1/2 h	−2/3 d + 1/2 h1
B2 ([1.2].2)		−1/3 d + 1/2 h2

three types of information: (1) genetic variation in a preharvest sprouting component assessed by measuring α-amylase, Hagberg falling number, or germination capacity directly, without reference to, or based on assumptions concerning, any physiological mechanisms, i.e., variation per se; (2) variation associated with a specific physiological system; and (3) variation associated with the segregation, or selection of, previously characterized, and sometimes apparently unrelated, genes. In cases where a rational association between the effects of such genes and PHS physiology is not evident, these correlations could be due to genetic linkage rather than a direct pleiotropic effect.

IV. THE PREMATURITY PREHARVEST SPROUTING PHENOTYPE

PHS damage caused before harvest ripeness is generally restricted to the accumulation of malt α-amylase, without visible germination, during the later stages of grain maturation. It has been noted as a genetic flaw in various U.K. winter wheats,[6,7,12,62] some South African varieties, and in triticales.[13,63] The distribution and levels of this α-amylase in ears of 'Maris Huntsman' from the 1985 U.K. harvest, in which season visible sprouting levels were low but the bread-making quality of the crop was seriously impaired, is shown in Figure 3a. Levels are highest in the proximal floret positions of the central spikelets within the ear, probably associated with the slower relative rate of the water loss from these grains.[8] In addition, large variation in the degree of α-amylase accumulation can also occur between ears within a crop.[62] Although reports of this source of α-amylase are few, the problem could be widespread but often overlooked. This is because, in analyses of harvest α-amylase levels, there is usually no way of assessing whether the enzyme was produced during grain maturation or post dormancy. In addition, crops in which α-amylase has developed during maturation may be prone to more rapid development of PHS symptoms following dormancy.[7] Development of prematurity α-amylase is highly dependent on weather conditions, and the problem is enhanced by any edaphic effects which extend the period to harvest ripeness, particularly during the period of water loss.[7]

A. Prematurity α-Amylase Production

Only a single study has been made of the breeding behavior of the high levels of α-amylase found in susceptible U.K. varieties at harvest.[12] Control appears to be simple; however, the continued presence of the trait in the Plant Breeding Institute breeding program some 20 years after the use of the Belgian variety 'Professeur Marchal', to which the problem can be traced, implies that its elimination may not be straightforward. Quite probably, the

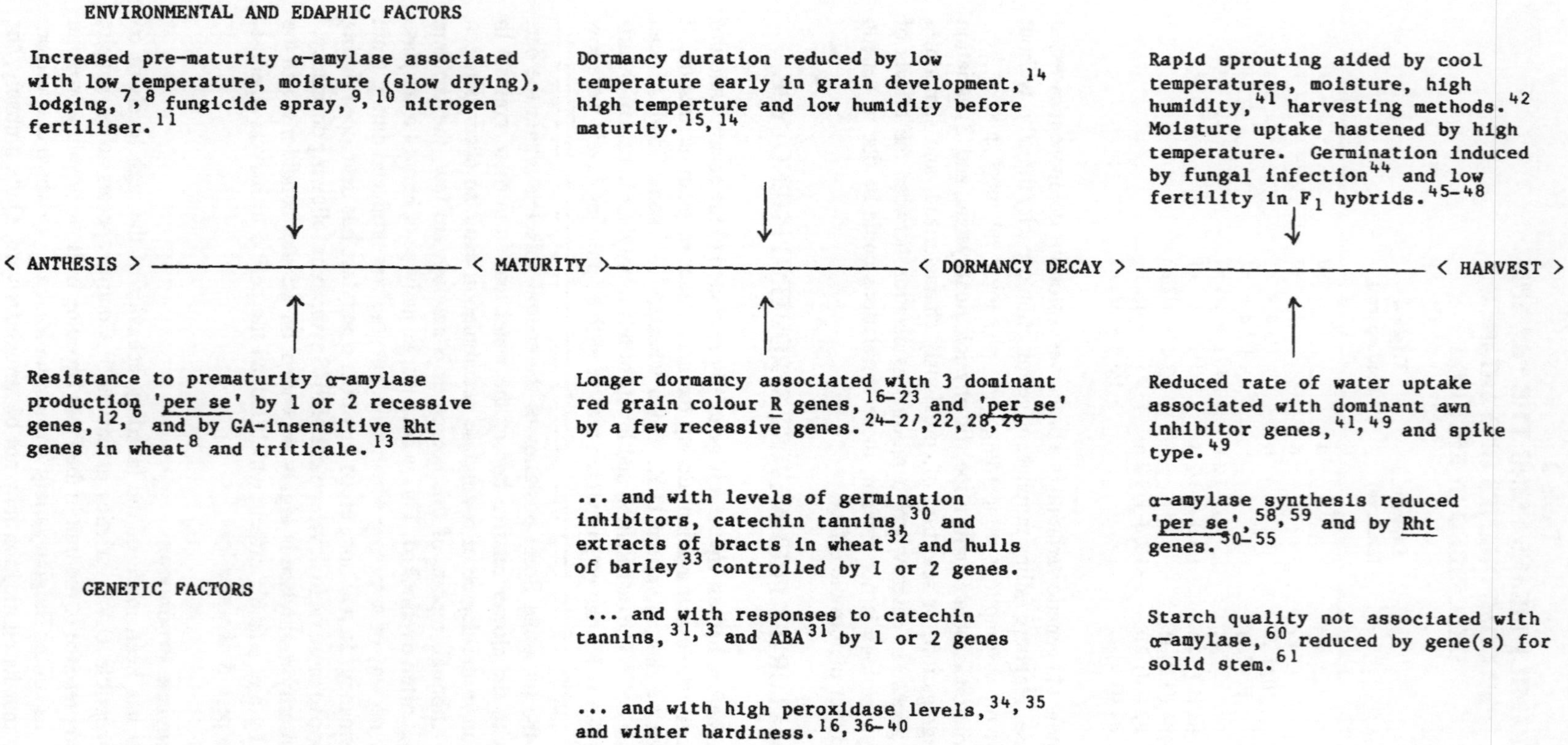

FIGURE 2. Summary of factors affecting the level of preharvest sprouting damage.

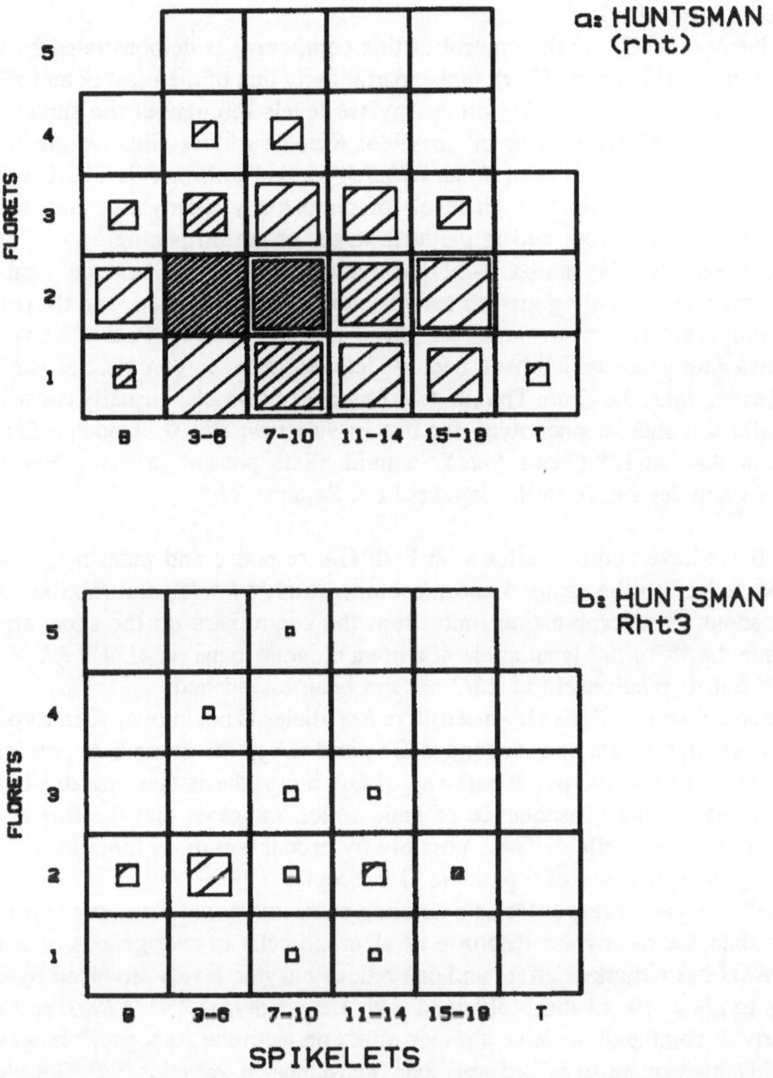

FIGURE 3. The presence of prematurity α-amylase in mature grains of (a) variety 'Maris Huntsman'; and (b) isogenic lines carrying the gibberellin-insensitive allele, *Rht3*. *Note:* The level of hatching in each cell indicates the relative quantity of germination α-amylase (α-AMY-1 and α-AMY-2), from high (<10 mU per grain) to low (<5 mU per grain), found in all grains at respective spikelets/floret positions. The proportion of each cell hatched indicates the proportion of grains at that position which have been triggered to produce germination α-amylase. (Based on Reference 8.)

sporadic occurrence of high α-amylase levels, about once every 5 years, has hindered progress. Further, the dominance relationship will allow the gene(s) to pass unnoticed in the heterozygous state. However, the possibility of maintenance via linkage to some beneficial trait in the original donor cannot be discarded.

High prematurity α-amylase "per se" — In a comparison of 29 F_4 lines from the cross 'Professeur Marchal' × 'Nord Desprez', a discontinuous distribution of 23 low and 6, high α-amylase families was observed.[22] This indicates control by a single, or at most two duplicate, recessive gene(s). No information is available concerning the site of action of the genes, but this is likely to be the aleurone (endosperm) or scutellum (embryo).

B. Gibberellin Insensitivity

The involvement of GA in the control of this component is demonstrated by the effects of the GA-insensitive dwarfing (*Rht*) genes, particularly that of the most potent allele, *Rht3*. Figure 3 shows the effects of *Rht3* on α-amylase levels throughout the spike in isogenic lines in the variety 'Maris Huntsman' in 1985; *Rht1*, a milder allele at the same locus, showed levels intermediate between *rht* and *Rht3*.[8] 'Maris Huntsman' is a U.K. winter wheat particularly prone to developing high levels of prematurity α-amylase, and in 1985 temperatures were extremely cool during the later stages of grain ripening.[3]

> **GA-insensitive *Rht* alleles** — Five *Rht* alleles, all of which can be identified by an insensitive seedling growth response to applied GA relative to that of tall *rht* genotypes,[64] are known in wheat. These are: *Rht1* and *Rht2*, the 'Norlin 10' semidwarfing genes which have been widely exploited in commercial varieties worldwide; *Rht3*, the 'Tom Thumb dwarfing gene,[65] which is usually considered to confer too short a phenotype for use in varieties; *Rht10*, a potent Chinese allele in Ai-bian-1;[66,67] and *Rht1S*, a mild allele present in many European varieties and derived from the Japanese cv. Saitama 27.[68]

- The alleles have additive effects on both GA response and plant height and are all carried as loci on the group 4 chromosomes. *Rht1*,[69] *Rht1S*, and *Rht3* are alleles at a locus about 13 recombination units from the centromere on the short arm of chromosome 4A.[69-71] *Rht2* is an allele at a homeologous locus on 4DS.[71] *Rht10* is also on 4DS,[67] but its relationship to *Rht2* has not been established.
- The mode of action of the GA-insensitive *Rht* alleles is not known. Genotypes carrying them have high levels of endogenous GA_1 and GA_{20}, the immediate precursor of Ga_1 in the early-13-hydroxylation pathway of GA biosynthesis that operates in vegetative plant tissue.[72] This consequence of gene action indicates that the *Rht* alleles cause nonutilization of available GA_1, possibly by production of an inhibitor or by directly reducing the efficiency of a putative GA receptor.
- The cell elongation responses in extending stem internodes are more sensitive to *Rht* action than the α-amylase response of aleurone cells to endogenous or applied GA. Only *Rht3* has a marked effect and can reduce enzyme levels produced by distal half-grains to about 1% of those obtained with *rht* genotypes.[50,51,56] *Rht1* and *Rht2*, particularly in combination, have a minor effect on aleurone response;[50] however, this is probably inadequate to afford any PHS resistance in varieties.[50,62] The aleurone responses of *Rht1S* and *Rht10* have not been reported. The aleurone GA-insensitive phenotype is entirely endosperm mediated and is allele-dose dependent. *Rht3 Rht3 rht* heterozygotes are similar in response to *Rht3* homozygotes, and *Rht3 rht rht* heterozygotes are almost entirely responsive.[51] A more complete review of the genetics, mechanism of action, and phenotropic effects of *Rht* genes is provided by Gale and Youssefian.[73]
- Dwarfing genes with analogous effects are known in barley (GA-ins),[74] rye, and maize (d8).[75,76]

Only the *Rht1* and *Rht3* have been assayed for their effects on prematurity α-amylase.[8,77] In isogenic lines, *Rht1* can reduce α-amylase levels, while *Rht3* virtually eliminates α-amylase at maturity. The effects of *Rht1*, and presumably *Rht2*, which in other respects has similar effects, are, however, unlikely to be of value for this reason in commercial varieties since other genetic sources of high prematurity α-amylase can be epistatic to the 'Norin 10' alleles, as testified by the propensity for some *Rht2* varieties, such as 'Fenman' and 'Hustler',[62] to display this kind of preharvest sprouting damage.

The *Rht3* allele, which could find use in F_1 hybrids, is also likely to be of only limited value.[78,79] By extrapolation from the results of allele dosage in the triploid endosperm during germination,[51] only one half of the F_2 grains on the F_1 crop, those with genotype *Rht3 Rht3 Rht3* and *Rht3 Rht3 rht*, will be protected.

V. THE DORMANCY PHENOTYPE

Variation in the duration of dormancy has been the most extensively studied of all the major components of genetic variation in sprouting damage. The most unequivocal genetic effect involves the association of red grain color with prolonged dormancy, although often, in studies of this relationship, it has been obvious that other genetic effects are also operating on dormancy duration.

Analyses of dormancy duration per se in crosses involving white wheats have uncovered valuable genetic variation, such as found in cvs. Kenya 321 and Ford.

Other dormancy associations with various ear and grain characters, in addition to grain color, have been noted, and include responses to catechin tannins, the presence of and response to various inhibitors, and the hull-less grain trait in barley.

Further associations with quantitative characters such as anthesis date, vernalization response, and winter hardiness have also been documented. With the exception of cold tolerance, which has, in some studies, been linked with variation in ABA, little further mention will be made of these latter associations. Apart from being of little value to the breeder, who will generally be restricted to breeding for specific vernalization response or anthesis date anyway, such correlations may well stem from environmental interactions, which can arise from shifting the grain development period within a variable temperature and percipitative weather pattern. This will be particularly true if the temperature sensitive "windows" which regulate the expression of dormancy are as short as those identified in barley by Reiner and Loch.[14] An additional reason for these associations could be simply, as pointed out by Weilenmann,[80] that earlier flowering is often reflected in earlier ripening, which in turn will tend to be associated with longer postharvest ripeness periods in the experimental field. The observations that earliness, among Australian varieties and segregating lines,[81] and spring habit are correlated with higher α-amylase levels are consistent with this hypothesis.[80]

A. Grain Color

The association between red grain color and dormancy has long been known and has long been a source of considerable annoyance to breeders of white wheats.[16,17,35] Nevertheless, discussion continues concerning the genetic and physiological basis of the relationship and the additive value of the several genes controlling grain color.

> **Red Grain Color**—It is generally acknowledged that red color is simply inherited and controlled by three independent, but probably homeologous, loci on chromosome arms 3AL, 3BL, and 3DL. The red alleles at these loci are symbolized, rather obtusely, as *R2*, *R3*, and *R1*, respectively.[82] However, at least two reports put the number of loci involved at more than three. Reitan[22] found five independent loci among eight varieties and Freed[20] found six. Interestingly, these six loci included two independent linked pairs of genes separated by 32% and 34% recombination, respectively, thereby raising the possibility that the long arms of the group 3 chromosomes carry pairs of duplicated loci.
>
> Several lists of wheat varieties cataloged for numbers of *R* alleles and available,[83,84] and more citations are provided in the *Catalogue of Gene Symbols for Wheat*.[82]
>
> Red color is dominant over white, and, because the pigment is formed in the testa layer of the pericarp, is inherited as the maternal genotype.

Much debate has surrounded the alternative hypotheses that the association is due either to a pleiotropic effect of the red color genes or independent "dormancy" genes which happen to be closely linked to the *R* loci. Although some biochemical and physical rationales linking red pericarp with a dominant reaction have been proposed,[19,85-87] a precise reason for pleiotropy has not been established. Nevertheless, the failure over many years to establish recombination between "linked" genes for the two phenotypes strongly indicates that pleiotropy is the reason for the association. Other circumstantial evidence leading to the same conclusion includes the involvement of genes which cause vivipary in the grain pigment biosynthesis pathway in maize, which is in turn linked with ABA synthesis.[88-91] Also, many of the varieties with exceptionally good dormancy, e.g., cvs. RL4137, Park, Thatcher, and Hilgendorf 61, carry *R* alleles on all three group 3 chromosomes. This in turn indicates that the effects of the *R* alleles are to some extent additive. It follows that, if these effects are due to genes linked to the *R* loci rather than pleiotropic effects of the *R* alleles themselves, then linked pairs of genes on all three chromosomes will be required to maintain a "linkage" hypothesis. From the above, it seems likely that a causal "pleiotropic" relationship exists.

A precise appraisal of additivity requires a method of establishing the number of *R* alleles in any genotype. Unfortunately, apart from extensive test crossing, there appears to be no direct way of assessing the genetic status of any red-grained line. Methods of quantifying the pigment in intact grains have often shown little association between intensity of color and gene dosage or dormancy.[22,23] Nevertheless, intra-allele additivity has been demonstrated by the fact that *R/r* heterozygotes display only partial dominance for dormancy.[92] Interallelic additivity has been demonstrated for comparisons of one- and two-allele and two- and three-allele genotypes.[20-22] Dominance, although not complete, is always in the direction of long dormancy duration for effects associated with the *R* alleles.

The question as to whether all the red alleles at each locus are equally effective appears only to have been addressed by Freed.[20] He concluded from pairwise comparisons of alleles that the *R* alleles at some loci confer more dormancy than others. However, this information is based on the assumption that there are six loci controlling red color rather than the more accepted view of only three, so the question must still be considered open. The only method available to approach this problem is to produce and test isogenic lines with the different *R* alleles isolated in a homogeneous genetic background. The answer, along with a clear understanding of the additive effects of different alleles, which will be best clarified by intercrossing isogenic lines carrying *R1*, *R2*, and *R3* and isolating two and three gene isogenics, is of the utmost importance to breeders of red-grained wheats.

B. Responses to Germination Inhibitors

1. Catechin Tannins

The role of catechins and catechin tannins (CT) as germination inhibitors have been demonstrated by Miyamoto and Everson.[30] Interestingly, these compounds are probable precursors of the red grain pigment, phlobaphene. The water-soluble colorless catechins and the yellow CTs are present in both white and red wheats, but at levels positively associated with grain color.[30]

Stoy and co-workers have shown that variation in sensitivity to germination inhibition by CT (Figure 4) is under simple genetic control and is associated with resistance to PHS.[3,31]

CT embryo response — Two results have been obtained which show sensitivity to CT to be under simple genetic control. Both involve the variety 'Snabbe' as the CT-insensitive (and sprouting susceptible) parent in crosses with (1) 'Breusted KV 6825 × 47 × Prins' (CT sensitive and PHS resistant) where F_3 analysis showed sensitivity to be controlled by two recessive genes;[31] and (2) U 67563, where F_1s, F_2s, and backcrosses showed susceptibility to CTs to be

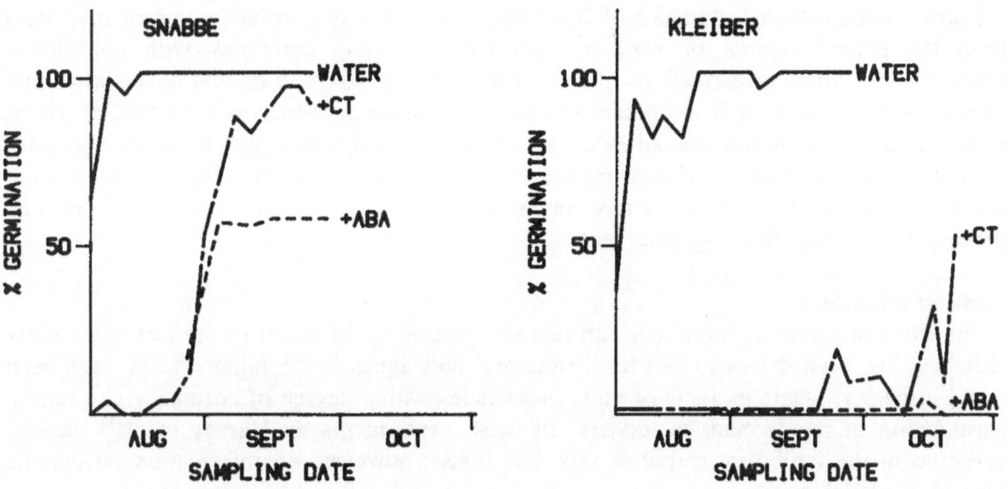

FIGURE 4. Differential responses to catechin tannins and abscisic acid in sprouting resistant, e.g., Kleiber, and sprouting susceptible, e.g., Snabbe, varieties. (After Stoy and Sundin.[31]).

due to a single dominant gene. The reciprocal F_1s and backcrosses investigated in this latter experiment showed that the CT-response was mediated, at least in part, by endosperm tissue.[3] A similar effect was noted in barley, where the sprouting-resistant CT-sensitive variety 'Helas' carried two duplicate recessive alleles for sensitivity.[3]

Unfortunately, no information is given in either of these reports concerning the *R* allele status of varieties or segregants examined. Nevertheless, the possibility exists that variation in the response to CT represents a further layer of control of dormancy, imposed on that controlled by the *R* alleles, i.e., where the level of inhibitory tannins, or, of course, metabolites of these compounds, is controlled by the pigment genes, while the degree of response to the inhibitors is under independent genetic control.

2. *Abscisic Acid*

Although evidence is lacking in wheat, in maize abscisic acid (ABA) is linked metabolically to the grain pigments, where the evidence now favors a carotenoid (C40) pathway, with violaxanthin as the immediate C40 precursor.[91]

Direct genetic evidence for the involvement of ABA, or responses to ABA, in the control of dormancy is not available. However, varietal variation in response clearly exists. Stoy and Sundin, in their investigations into the genetics of sensitivity to CT,[31] also observed the effects of ABA on germinability of grains sampled at various times following HR for four varieties (Figure 4). The results were similar to those obtained with CT for the same genotypes, in that ABA was most effective in the inhibition of germination of isolated embryos and intact grains in the PHS resistant varieties. The PHS susceptible varieties tested, 'Snabbe' and 'Amy', were relatively insensitive to ABA.

The levels of ABA can be manipulated by selection in wheat,[93] although not as dramatically as is possible by mutation in diploid species. The fact that ABA-less mutants in maize,[94] arabidopsis,[95] tomato, and potato all display vivipary[91,96] (i.e., embryo development is not arrested and continues through to germination without entering a dormant phase), is indicative that high ABA levels, or high responsiveness of embryo tissues to ABA, may be associaed with longer dormancy duration in wheat. An ABA-less mutant, *albostrans*, exists in barley;[97] however, its dormancy phenotype has not been reported.

Further circumstantial evidence of involvement of ABA in dormancy control may stem from the several reports of varietal associations of long dormancy with cold tolerance.[16,36-40,117] Some investigations of ABA levels during exposure to cold have shown that accumulation is more rapid in tolerant varieties than in the cold-susceptible varieties. However, the evidence is not unequivocal, since Quarrie and Cloutier,[40] using the varieties 'Kharkov', as an example of cold-resistant genotype, and 'Cappelle-Desprez' as a cold-sensitive example, failed to show any consistent differences in ABA levels, either between genotypes or during freezing treatment.

3. Bract Inhibitors

An effect of a soluble inhibitory substance(s) present in the bracts of the German variety 'Kleiber' has been demonstrated by Dererá and colleagues.[32,98] Similar effects have been noted in barley, where extracts of hulls of varieties with a degree of dormancy can inhibit germination in nondormant genotypes. In these experiments by Harvey et al.[33] varietal variation in the inhibitory response was also noted; however, no information on genetic control is available.

> **Presence of bract inhibitors** — F_3 and F_4 families from the cross 'Kleiber' (red grain, inhibitor present, preharvest-sprouting resistant) × 'Gamut' (white grain, no inhibitors, preharvest-sprouting susceptible) were examined.[32] The test involved germinating corresponding grain samples in the presence of milled bracts and water to assess inhibition levels. Active inhibition was found in 15% and 24% of the families in the two generations, indicating simple control by one or two recessive genes. Reciprocal differences, which would indicate cytoplasmic control, were not observed.

The relationship of these inhibitors, if any, to the pigment-related inhibition discussed above is not known because the design of the experiment would not distinguish between the presence of, or response to, inhibitory substances and, furthermore, the grain color of the segregants was not reported. However, since these experiments involve 'Kleiber', the same variety as that shown by Stoy and Sundin to be highly sensitive to both CT and ABA,[31] it is clearly possible that the "bract inhibitors" are related to the inhibitors discussed above.

4. Is There an Inhibition-Response Mechanism?

Further evidence of genetic control of responsivity of germinating grain to some product associated with red grain color is provided by experiments by Noll et al.,[99] where reciprocal F_1 grain from crosses involving the variety 'RL4137' were tested for germination over an 8-day period. One of these crosses involved cv. Neepawa as the second parent (Figure 5). This variety, like 'RL4137', has *R* alleles at all three loci, so all the F_1 grain from the cross will have been homozygous for the *R* alleles and, therefore, any differences observed will be due to other factors. Both F_1s showed intermediate germination rates. Furthermore, the two reciprocals differed, indicating that the control was mediated, at least in part, by genes expressed in the endosperm, just as was found for the genes controlling CT response.

A two-tiered "inhibitor-response" genetic mechanism associated with grain pigments and precursors, or metabolically related compounds, is consistent with the common observation that much more variability for dormancy duration exists among red wheats than among white varieties. It is also consistent with the evidence that red varieties with extreme dormancy, such as 'RL4137', may derive their phenotype from both genetic sources. Indeed, De Pauw and McCaig showed that usable variation for long dormancy could be isolated in white grain segregants from a cross involving 'RL4137' with '7722', a white-seeded wheat.[29] It remains to be seen whether the same dormancy characteristics could be achieved by selecting for high CT or ABA levels and for a sensitive response to such germination inhibitors.

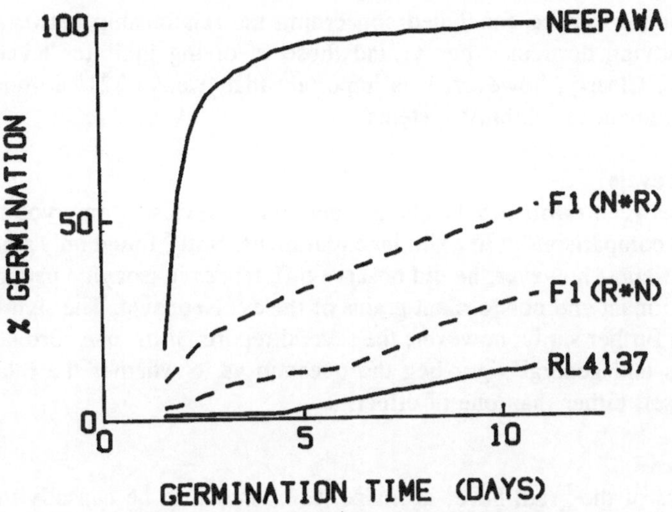

FIGURE 5. Germination rates in reciprocal F_1 hybrids involving cvs. RL4137 and Neepawa, both of which carry three *R* alleles. (After Noll et al.[99]).

C. Other Sources of Dormancy

Direct screening of white-grained varieties has identified a number of varieties with relatively long dormancy, although considerably less than red-grained standards such as RL4137. The most studied is probably 'Kenya 321'; however, in a screen of 3000 other genotypes, Mares identified a number of genotypes with longer dormancy than 'Kenya 321', such as 'Hellas', 'AUS 1408', and 'AUS 1490'.[100]

Other studies of the genetics of dormancy duration per se that have been widely quoted are not discussed below because of difficulties in interpretation because red-grained parents have been used in crosses and the segregating effects of the *R* alleles have not been accounted for in the progenies examined.[25,26,101]

Dormancy duration is a difficult trait to quantify precisely, because it is generally impractical to sample germination at postharvest intervals of less than 1 week. Also, of course, it is difficult to predict the range of dormancy duration expected in any particular season. Problems such as these are almost certainly the reason for the paucity of information concerning other genetic systems affecting dormancy.

'Kenya 321' Dormancy—One study by Bhatt and Derera provides convincing evidence of major gene control.[27]

1. 'Kenya 321' — Dormancy, measured as the proportion of germinable grain after cold storage at harvest ripeness, was assessed in F_1s, F_2s, and backcrosses of the crosses 'Kenya 321' × the white, nondormant cvs. 'Shortim' and 'Gamut'. Long dormancy was found to be recessive, with no maternal component. The ratios were compatible with control by two duplicate genes.[27]
2. 'Ford' — An identical analysis of the cross 'Shortim' × 'Ford' showed a similar result.[28]

The magnitude of the effects of genes such as these relative to those of the *R* alleles is difficult to assess. In some studies involving progenies of red-grained × white-grained parents, the *R* alleles account for virtually all the dormancy variation.[18] However, in a study by Reitan[22] of a diallel cross involving eight parents, it was found that, while the difference between red- and white-grained phenotypes was greatest, recessive genes for long dormancy duration were of greater effect than the differences between one and two *R* alleles.

Of course, nothing can be concluded concerning the relationship between mechanisms identified by studying dormancy per se and those involving inhibitor levels or inhibitor response systems. Clearly, however, it is important that 'Kenya 321' dormancy be investigated for involvement of inhibitor systems.

D. Peroxidase Levels

Correlations between dormancy levels and peroxidase levels in embryos have been observed in varietal comparisons.[34] In a similar experiment, Noll[35] found only small differences between three varieties; however, he did observe differences in isozyme patterns for anionic peroxidases of dormant and nondormant grains of the cv. Neepawa. The significance of this result must await further study; however, the several reports of inverse correlations between peroxidase levels and general vigor beg the question as to whether the relationship with dormancy is causal, rather than one of effect.

E. Fungal Diseases

Fungal diseases of the grain have, in maize, been shown to be causally involved in the induction of PHS. *Diplodia maedis* and *Fusarium moniliforme* infections have,[43,102] in field and laboratory inoculation of intact cobs, been shown to induce germination. The presumed mechanism is that the pathogens either produce growth-regulating compounds or cause endogenous hormone imbalance in the host, which in turn induces germination.

Fungi such as *Fusaria* (scab) and *Alternaria* (black point) can attack wheat grains and could have similar detrimental effects on PHS phenotype. Some strains of *Fusarium graminearum (Gibberella zeae)* can attack both wheat and maize and have, indeed, been reported to grow into the wheat endosperm and cause the initiation of starch breakdown.[44] No information on the genetics of resistance to *Fusaria* spp. is available.

Recent reports indicate that fungicidal spray treatments can themselves lower Hagberg falling number values.[9,10] This effect is probably via prematurity α-amylase production rather than by germination initiation.

VI. THE POSTDORMANCY PHENOTYPE

Even after dormancy ceases to be a protection against preharvest sprouting, other environmental and genetic factors will determine the degree of damage attained prior to harvest. Moisture is obviously required, from untimely rain or even dew on cold nights. The maintenance of high humidity, such as in a lodged crop or, in particularly damp conditions, in an extremely short crop,[11] will similarly hasten PHS. Thereafter, the genetic components of PHS damage at this time include those which affects: (1) the rate of water uptake into the ear and the grain, such as spike morphology and attitude; (2) the potential for rapid germination, which may be enhanced by high temperatures; and (3) the rate of production of germination enzymes, such as α-amylase, which will be affected both by the rate of production of GA in the germinating grain and by variation in response to endogenous GA.[103]

A. Water Uptake

Variation has been demonstrated in the ability of grains in intact ears under simulated rain conditions,[104] and in detached grains,[45,105] to take up water. No information is available on the genetic control of this character; however, King and Richards[49] did show that some of the variation was not associated with pericarp characteristics, including grain color.

Spike morphology has been studied in relation to rate of wetting in a sample of Australian varieties and series of near-isogenic lines differing in genes controlling awning, pubescence, glaucousness, and head type.[41,49] The presence of awns was clearly implicated. Awned genotypes showed higher, by 30%, rates of wetting and enhanced, by 40%, levels of preharvest sprouting. The awned isogenic lines also displayed reduced water loss under drying conditions.

Awning — The presence of awns in wheat is controlled mainly by three dominant awn-inhibitors: *Hd* (hooded) on chromosome 4BS, *B1* (tipped 1) on 5AL, and *B2* (tipped 2) on 6BL. The presence of any two of these alleles is required for a completely awnless phenotype. The modifying effects of recessive awn promoters are also documented.[82]

The compact ear of club wheat, relative to the more lax spike of common wheat, also increased the rate of water uptake.[49] presumably through increased water retention between the compressed spikelets.

Club ear type[82] — This phenotype is controlled by a single dominant gene, *C*, carried on 2Bβ. Common wheats have the genotype *c, Q* (nonspeltoid ear), or *S1* (nonspherococcum type), while club wheats carry *C Q S1*.[82]

B. Rates of Germination

Varietal variation exists in rates of germination of nondormant grains. Mares identified the variety 'Saberberg' as having slow germination not associated with dormancy or water uptake rate.[100] However, no information on genetic control is available.

C. Rates of Production of Germinative Enzymes

Investigations of varietal differences in rates of production of germinating hydrolases have uncovered effects which correlate with levels of resistance to PHS.[58,59,105] The physiological and genetic bases of these differences are not known, but could relate to differences in rates of GA production by the embryo or differences in sensitivity of the scutellum or aleurone cells to GA.

The most dramatic reported differences in α-amylase production, and other GA-induced germination enzymes, are again associated with the GA-insensitive dwarfing genes, in particular *Rht3*, the 'Tom Thumb' gene. In GA-treated de-embryonated endosperm half-grains, α-amylase levels in *Rht3* homozygotes can be only 1/100th the levels in comparable tall controls.[50] In intact grains, either under sprouting conditions in the field or under laboratory germination conditions, the difference is reduced to about 1/10th.[54] The reason for this difference may be that the insensitivity imparted by *Rht3* is more effective in aleurone tissue than in the scutellum, which is an important source of α-amylase in the early stages of germination.[106] Also, it is clear that the reduction in α-amylase is not adequate to impede the rate of germination in *Rht3* genotypes. The wide range of visible sprouting among groups of random F_4 tall, *rht*, and dwarf, *Rht3*, homozygotes observed by Flintham and Gale was unaffected by genotype;[54] however, the *Rht3* lines had consistently less α-amylase, by almost 80%.

The possibility exists that the GA-insensitive dwarfing gene locus on 4AS, at which *Rht3* is an allele, is particularly variable. Already, two other alleles, *Rht1* and *Rht1S*, both with milder effects than *Rht3*, are known at this locus. The studies of the effects of *Rht3* in U.K. winter wheat backgrounds have utilized 'Minister Dwarf' as the dwarfing gene source.[50,56,69] This gene, via 'Tom Thumb', probably traces back to a single British source, 'Hybrid Carter G'.[107] Other studies, particularly those involving *Rht3* in crosses with Australian spring wheats,[52,53,55,57,106] have utilized the CIMMYT variety 'Tordo' as the source of the dwarfing gene. Although under some conditions 'Tordo' is as GA insensitive as European *Rht3* genotypes,[52] this variety can occasionally give very variable results.[55,108] The CIMMYT *Rht3* gene is reported to derive from a Tibetan source, and the possibility remains that it differs from the 'Minister Dwarf' source.[109] No answer to this question can be made without the development of appropriate isogenic lines. However, if the *Rht* allelic forms are based on differences in gene copy number, then further alleles intermediate in effect between *Rht1S* and *Rht3* may exist.

D. Starch Quality

There are aspects of starch quality, under genetic control as judged by varietal differences,[60,110,111] which can affect Hagberg falling number values, although not α-amylase-levels, in flour. Moss[110] has investigated differences in noodle quality and found that, particularly at low α-amylase levels, Hagberg falling number and viscosity can be related to starch granule size and amylose content of the flour, rather than α-amylase levels, in a comparison of four varieties. High viscosity and low amylase are required for boiled Japanese (Udon) noodles, while the opposite is the case for Chinese (Ramen) noodles.

Analyses by Chojecki et al.[112] showed clearly that starch granule size is under genetic control and, in an analysis of chromosomal effects, was positively associated with genetic effects on grain size.

Another possible effect on starch quality, lowering the Hagberg falling number for a given α-amylase value, has been noted by McCaig and DePauw[61] to be associated with the character "solid-stem", used in breeding saw fly (*Cephus linctus*) resistant wheats in Canada. Pedigree analyses of the varieties 'Leader' and 'Chester' trace the low Hagberg falling number characteristics of these varieties back to 'S615', the source of solid-stem. The genetics of solid-stem are not entirely characterized, but analyses of monosomics in 'S615' by Larson and MacDonald[113] indicate major gene control on the group 3 chromosomes.

VII. α-AMYLASE

All the discussion above concerns sources of genetic variation in components of the preharvest sprouting syndrome, which causes the induction of α-amylase production and germination. The α-amylase structural genes themselves should provide some possibilities for manipulation to reduce weather damage. For example, less efficient genes would be beneficial in this respect and would probably not adversely affect germination capacity. One inference to be made from the ability of *Rht3* dwarfs to germinate well is that the high levels of α-amylase produced in conventional wheats represent production well above the minimum necessary for grain mobilization. Similarly, mutant alleles which produce an enzyme with a thermal inactivation point below 65°C, the gelatinization temperature of starch, would instantly solve most of the problems that α-amylase causes in the commercial breadmaking process. Unfortunately, the organization and complexity of the α-amylase multi-gene family in wheat would appear to limit our hopes for these types of solution to the preharvest sprouting problem.

The enzymes of significance in bread-making quality are the α-AMY-1 or "malt" isozymes, rather than the α-AMY-2 or "green" isozymes that are also produced during germination.[114] The significance of a third enzyme, α-AMY-3 is not known.[114,115] Not only are the α-AMY-1 isozymes produced by three loci, but the loci themselves are complex, comprising several linked genes which produce proteins of similar effect, but which are detectable by their altered electrophoretic patterns. To produce any significant effect on α-AMY-1 quantity or quality, the breeder would have to manipulate or inactivate almost all of these independent genes.

Before describing our present appreciation of the genetics of α-amylase, a few notes on the method of naming protein genes in wheat, outlined in detail by Hart and Gale,[116] may be of value, because there are several nomenclatures in use for the enzyme and its isozymes. Where evidence of a set of loci controlling production of homologous proteins are found over two or more genomes in the Triticeae, the set is defined as (e.g.) *α-Amy-1*. Subsequently discovered sets of genes with similar functions are given sequential set numbers, e.g., *α-Amy-2*. Individual loci, which may comprise a cluster of closely linked gene copies, as is the case for α-amylase genes, are named with their genomic symbol, i.e., *α-Amy-A1* or *α-Amy-R1* (for the homologous gene in rye). Alleles, observed as a different pattern of isozymes produced by any single locus, are identified as *α-Amy-A1a, α-Amy-A1b*, etc. The "a" allele is usually reserved for the genotype of the variety 'Chinese Spring', which is the background genotype of many of the genetic stocks used in wheat genetical analysis.

Unfortunately, much of the work on α-amylase predates the elucidation of the formal genetics of the genes involved, and indeed the method of symbolizing these genes. The terms "malt" and "green" were first used by Swedish workers, and the group at the Canadian Grain Commission coined the terms GI, GII, and CIII based on pI ranges of amylase isozymes.[117,118] Gene products, in wheat genetics terminology, are named as the genes, but in upper case, e.g., α-AMY-1. Several nomenclatures are also in current use for the barley enzymes. However, in barley genetics, the loci are named in line with the wheat system,[119] *α-Amy 1* and *α-Amy 2*, and may, where necessary, be discriminated from the wheat loci by inclusion of the genome symbol, e.g., *α-Amy-H1*.

A. α-Amylase Genes

Three species of α-amylase, each under separate genetic control have been described.

1. α-AMY-1

Known also as "malt", high pI, Group 1, and GIII, α-AMY-1 is produced after induction by GA during germination but also, occasionally, during the latter stages of grain maturation.[118,120] Three compound loci on the long arms of chromosomes 6A, 6B, and 6D,[121] located about 20 cm from the centromere,[122] code for the 17 or so isozymes with pI between 6.5 and 8.0. Each locus controls production of more than one isozyme. Genotypes nullisomic for chromosome 6D (and therefore lacking the *α-Amy-D1* locus), for instance, show the loss of seven different isozymes.[123]

The estimates of gene number at the *α-Amy-1* compound loci in wheat are similar to the number of isozymes observed. These range from nine, as seen from in vitro translation of the *α-Amy-1* mRNAs, to 14,[124] as shown by the number of restriction fragments which hybridize to *α-Amy-1* cDNA probes.[125] A range of alleles at these loci, as distinguished by differences in number or mobility of isozyme bands, has been demonstrated. *α-Amy-B1* is the most variable, with eight alleles identified among 200 varieties.[126] One report indicates the presence of a further duplicated locus on chromosome 6B (α-Amy 1), located 21 recombination units distal to α-Amy-B1;[122] however, subsequent attempts to show recombination among the α-amylase genes on 6B have failed.[126]

2. α-AMY-2

Known also as "green", low pI, Group 2,[120] and GI and GII,[118] α-AMY-2 is produced in the endosperm during germination and is also found in the pericarp early in grain development. Only a few isozymes are found in the pericarp relative to the 16 in GA-induced half-grains, but genetic evidence from analysis of nullisomic-tetrasomic stocks and allelic differences at the *α-Amy-2* loci indicate the enzymes in both tissues are controlled by the same genes.[127] α-AMY-2 production in the pericarp is not GA-induced.[128]

A similar number of gene copies and isozyme forms are found for α-AMY-2 as for α-AMY-1. Three compound loci on 7AL, and 7BL, and 7DL,[121] located about six recombination units from the centromere,[123] control production of all the isozymes.

Allelic variation is rarer for the *α-Amy-2* set than *α-Amy-1*, but variants at *α-Amy-B1* and *α-Amy-D1* have been described.[126]

3. α-AMY-3

Recently two independent reports have indicated the presence of a third form of α-amylase in wheat. Daussant describes a group of isozymes,[129] with very high pI (>pH10), produced only in the outer pericarp layer during early grain development. Baulcombe et al.[115] have found a third species of α-amylase gene which produces mRNA only during early development. Expression, unlike the germination enzymes, is not GA induced, and only a few copies of the *α-Amy-3* genes are present on the group 5 chromosomes. Plainly, these two reports could refer to the same third α-amylase species; however, confirmation of this must be awaited.

Analogous genes, carried at homoeologous loci throughout the Triticeae and α-amylase loci, of similar function and probably of similar complexity to those in bread wheat, can be demonstrated on the long arms of chromosomes homoeologous to wheat groups 6 and 7 in barley, rye, and their wild relatives.[130] The presence of *α-Amy-3* in other species has not yet been reported.

B. α-Amylase Inhibitors

Two types of protein inhibitors of α-amylase are present in cereal grains and have been reviewed by Silano.[147] The most abundant group, first reported by Kneen and Sandstedt, which comprise about 1% of total protein of mature grains,[131] are members of three families of albumin proteins which are active against mammalian and insect amylases, but do not inhibit cereal amylases. The levels of these proteins drop rapidly during germination and only 5% of the original level remains after 2 days germination. This is consistent with their role being associated with a rapid supply of nitrogen and amino acids for the embryo following proteolytic breakdown in the endosperm.

The second type of inhibitor, which is active against native α-amylases, is of more interest in connection with preharvest sprouting. Evidence of multiple forms of endogenous α-amylase inhibitory proteins was first provided by Warchalewski.[132] Levels of these proteins increase during germination, and it is possible that their function may be to control α-amylase activity at this time; however, there is at present no evidence which indicates this role.

The wheat amylase-subtilisin inhibitor, WASI, and an analogous protein,[133] BASI,[134,135] in barley grains comprise a third category of inhibitor. BASI, first purified from barley grains by Yoshikawa et al.[136] as an inhibitor of bacterial proteases, appears to be double headed, and to specifically inhibit α-AMY-1 by forming complexes with malt isozymes at sites other than those directed at subtilisin.[135] WASI and BASI levels are not increased during germination.

> **Exogenous α-Amylase Inhibitors**—Albumin proteins, now known to have activity against both insect and mammalian amylases with molecular weights around 24,000, known as the 0.19 family, have been shown to be controlled by genes on chromosomes 3A, 3B, and the short arm of 3D.[137-139] A major inhibitory fraction, purified as Inh I, was located on 3Bs.[140] The 0.19 inhibitor (also termed Inh III),[140] and the main component of this family, is among the proteins controlled by genes on 3D.[140] The protein has been found in *Aegilops squarrosa*,[141] the donor the D genome, but not in durum ($2n = 4x$ (AABB) $= 28$) wheats, and has been employed as a test for adulteration of durum products with bread-wheat flour.[142]
>
> Albumins with molecular weights of around 12,000, known as the 0.28 family, which have been shown to have activity only against insect amylases, are controlled by genes on chromosome 6B and 6DS.[139] A major inhibitor fraction, purified as Inh II,[140] has been located on 6DS.
>
> **Endogenous α-Amylase Inhibitors**—The genetics of the albumins with molecular weights around 19,500, that have inhibitory effects on native cereal amylases, have not been studied.
>
> The chromosomal location of WASI is likely to be the group 2 chromosomes. Hejgaard and colleagues showed the gene controlling BASI, *Asi1*, to be located on 2H (≡ barley chromosome 2).[143] A single WASI protein band reacted with

the monospecific antibodies used in the genetic analysis of the wheat-barley chromosome addition lines employed in the genetic analysis. This indicates that control in wheat is either by a single gene, probably on the group 2 homoeologues, or more likely, by more than one gene with isoelectrophoretically similar products.

Varietal variation in detectable levels of endogenous inhibitors has been observed, and the monospecific antibodies available could be used as a selection tool to manipulate and evaluate these inhibitors as a potential means of reducing starch damage by α-amylase in sprouted grain.[144]

VIII. F_1 HYBRID WHEAT

Following the recent availability of a new generation of efficient gametocides, now known as chemical hybridizing agents (CHAs), the potential of F_1 hybrid wheat is being widely reassessed. Although, of course, the breadmaking quality of the F_1 grain itself is unimportant, its necessarily high price will dictate that it must be sound. Earlier work utilizing genetic male sterility, usually based on cytoplasm of *Triticum timopheevi*, indicated that severe PHS could often be observed in F_1 production plots.

Several studies showed that the degree of preharvest sprouting found in F_1 crossing plots was closely related to levels of seed set attained.[45-48] Initially this was attributed to cytoplasmic-nuclear interactions;[45] however, the similar relationship found with Ethephon-induced sterility, indicated a nongenetic cause.[47] In recent experiments at the Plant Breeding Institute with the Shell chemical WL 84811, severe sprouting was observed on one occasion in back tillers where seed set is lowest.[145] However, since fertility levels of this order will not be commercially viable, the indications are that, when hybrid grain production can be achieved economically, preharvest sprouting will not be an associated problem.

Otherwise the F_1 breeder is in the happy position that he can exploit all dominant maternally inherited traits. The simple expedient of ensuring that at least one, and preferably more, *R* allele(s) is carried by either parent will be of obvious value. Similarly, if awns are not required for other purposes, only one of the parents need be awnless to ensure that the adverse effects of awning on preharvest sprouting are not present in the F_1 crop.

Other traits, expressed in the endosperm or embryo tissues, will, of course, segregate freely in the F_2 grain on the F_1 crop. Single gene dominant effects, such as a strong response to inhibitors, may be used to give 3:1 cover in the F_1 crop. However, the utilization of recessive traits such as have been found for low prematurity α-amylase and for the 'Kenya 321' type of dormancy will require careful selection of both parents.

F_1s, in which the traits of high prematurity α-amylase, from 'Maris Huntsman', and GA-insensitivity, from *Rht3*, have been combined, have been assessed in a season where prematurity α-amylase was generally high.[78] As expected, the dose-dependent expression of *Rht3* in the triploid endosperm ensured that α-amylase levels in the F_1s fell midway between the parents. Although, in this experiment, some *Rht3/rht* hybrids had α-amylase levels which were still too high for quality purposes, it was important to demonstrate that heterotic levels of the enzyme were not produced, because this genotype shows promise for use in commercial F_1s.[79]

IX. SUMMARY, AREAS OF IGNORANCE

This overview of the genetics of preharvest sprouting component characters indicates clearly that more knowledge of genetic control of the many varietal sources identified is required before rational breeding strategies to produce bread and noodle-making quality wheats with consistent resistance to preharvest sprouting can be devised for all locations. This is, of course, especially true for the breeders of white-grained wheats in Australia and parts of North America.

Also, although the importance of prematurity α-amylase has only recently become recognized as a component of preharvest sprouting, it is likely that this is a more widespread problem than the few reports of its occurrence suggest. It is probable that many of the reported analyses of α-amylase levels or Hagberg falling number, carried out ostensibly to assay dormancy, will have confounded prematurity and postdormancy sources of the enzyme. Future screens of potential breeding material should include separate assays of the enzyme in harvest-ripe material and at later dates. Our limited knowledge of the underlying genetic basis of the high prematurity α-amylase syndrome represents a major omission in our armory. The single genetic analysis reported indicates a simple mode of inheritance, and the eradication of one or two dominant alleles from breeding programs should be a simple matter. Nevertheless the widespread occurrence of the problem, and indeed its continued presence in programs such as that at the Plant Breeding Institute, where selection for quality has been the highest priority, may indicate the involvement of more genes, possibly involving several physiological systems, or even a pleiotropic or linkage association with other characters of beneficial agronomic significance.

The value of the red grain color, *R*, alleles has been recognized for more than 50 years; nevertheless major gaps in our knowledge still impede the optimal use of these genes. Specifically, information is required as to whether all the *R* alleles are equally effective in prolonging dormancy and whether all the alleles are additive in their effect. Also, confirmation is required as to whether the effect actually has two components, i.e., the level of pigment, or related compounds, in the testa and the sensitivity of the endosperm or the embryo to these compounds. Pending clarification of these points, efforts should be made to develop effective assays for the number of *R* alleles in segregating progenies. The response system, which may be genetically simple, should be more closely defined and selection methods may need to be developed. In addition, although the numbers of *R* alleles in many varieties have been cataloged, the number of varieties for which the identity of the R alleles is known is limited. A comprehensive survey, at present only possible by test crossing, would be of great value to breeders. Finally, we are still lacking the isogenic stocks with *R* alleles at each locus which are needed for precise evaluation of the red color genes. Such lines must be produced by backcrossing into defined genetic backgrounds, rather than the easier alternative method of holding genes heterozygous through a number of selfing generations and finally selecting isogenic homozygous pairs of genotypes carrying different alleles. These latter "isogenic pairs", which are useful for comparing the effects of alternative alleles, do not permit alleles to be combined for the assay of additive effects in isogenic backgrounds.

Varietal sources of true long-duration dormancy, quite independent of the effects of *R* alleles, plainly are available. However, the lack of knowledge of the physiological mechanism underpinning the dormancy characteristics of these varieties is restricting the advance of our understanding of the range and complexity of genetic systems involved. The dormancy phenotype is time consuming to quantify with present methods, and more efficient selection systems would be welcomed by breeders. Indeed, "dormancy" genes are ideal candidates for "tagging" by markers such as the restriction fragment length polymorphisms (RFLPs) presently being employed to achieve a total genetic map of the wheat genome. RFLPs will have an advantage over conventional morphological markers in that they will usually be "breeder-transparent", i.e., will have no effects, other than via genetic linkages, on the range of yield and quality characters that are already being selected.

The postdormancy components of the total preharvest sprouting syndrome have as yet been inadequately studied to allow the development of a reliable genetic strategy. The effect of awns on wetting rate does appear, however, to be conclusive and the advantages of awnless genotypes for areas where preharvest sprouting is a problem can now be weighed against the possible yield benefits reported for awned types.[146]

Finally, the α-amylase genes themselves appear, at present, to provide little promise for manipulation or modification by conventional breeding techniques. Nevertheless, they are already so well characterized at the molecular level that, when the genetic engineering techniques of transformation and regeneration of the small grain cereals become possible, their directed removal and modification or partial inactivation, could provide a spectacular solution to a major part of the sprouting problem in wheat.

REFERENCES

1. **Miller, T. E.,** Evolution and systematics in wheat breeding and genetics, Lupton, F. G. H., Ed., *Wheat Breeding; Its Scientific Basis,* Chapman & Hall, London, 1987, 1.
2. **Hart, G. E. and Gale, M. D.,** Biochemical/molecular loci of hexaploid wheat (*T. aestivum,* 2n = 42, genomes AABBDD), in *Genetic Maps 1987. A Compilation of Linkage and Restriction Maps of Genetically Studied Organisms,* Vol. 4, Cold Spring Harbor Laboratory, Cold Spring Harbor, N.Y., O'Brien, S. J., Ed., Cold Spring Harbor, N.Y., 1987, 678.
3. **Stoy, V. and Olsen, O.-A.,** Inheritance of a factor affecting the response to germination inhibitors in excised wheat embryos, in Proc. 2nd Int. Symp. Preharvest Sprouting in Cereals, Gale, M. D. and Stoy, V., Eds., *Cereal Res. Commun.,* 8, 203, 1980.
4. **Fisher, R. A., Immer, R. F., and Tedin, O.,** The genetical interpretation of statistics of the third degree in the study of quantitative inheritance, *Genetics,* 17, 107, 1932.
5. **Mather, K. and Jinks, J. L.,** *Biometrical Genetics, The Study of Continuous Variation,* 2nd ed., Chapman and Hall, London, 1971.
6. **Gale, M. D.,** High α-amylase — breeding and genetical aspects of the problem, Proc. 1st Int. Symp. Preharvest Sprouting in Cereals, Stoy, V. and Derera, N. F., Eds., *Cereal Res. Commun.,* 4, 231, 1976.
7. **Gale, M. D., Flintham, J. E., and Arthur, E. D.,** Alpha-amylase production in the late stages of grain development — an early sprouting damage risk period?, in *Proc. 3rd Int. Symp. Preharvest Sprouting in Cereals,* Kruger, J. E. and LaBerge, D. E., Eds., Westview Press, Boulder, Colo., 1983, 29.
8. **Gale, M. D., Salter, A. M., and Lenton, J. R.,** The induction of germination alpha-amylase during wheat grain development in unfavourable weather conditions, *Proc. 4th Int. Symp. Preharvest Sprouting in Cereals,* Mares, D. J., Ed., Westview Press, Boulder, Colo., 1987, 273.
9. **Morgan, A. G.,** Observations on the effects of fungicide and grain storage on Hagberg Falling Number, in Proc. Symp. on Some Aspects of Wheat Milling Quality, National Institute of Agricultural Botany, Cambridge, May 15, 1986.
10. **Davis, W. P.,** Responses of Hagberg Falling Number to nitrogen and late-season fungicides, in Proc. Symp. on Some Aspects of Wheat Milling Quality, National Institute of Agricultural Botany, Cambridge, May 15, 1986.
11. **Ciha, A. J. and Goldstein, W. A.,** Effects of fertility and rain simulation during grain fill on protein content, starch quality and alpha-amylase activity in winter wheat, in *Proc. 3rd Int. Symp. Preharvest Sprouting in Cereals,* Kruger, J. E. and LaBerge, D. E., Eds., Westview Press, Boulder, Colo., 1983, 72.
12. **Bingham, J. and Whitmore, E. T.,** Varietal differences in wheat in resistance to germination in the ear and α-amylase content of hte grain, *J. Agric. Sci.,* 66, 197, 1966.
13. **Chojnacki, J., Brykczyinski, and Tymieniecka, E.,** Preliminary information on sprouting in triticale, in Proc. 1st Int. Symp. Preharvest Sprouting in Cereals, Stoy, V. and Derera, N. F., Eds., *Cereal Res. Commun.,* 4, 111, 1976.
14. **Reiner, L. and Loch, V.,** Forecasting dormancy in barley — ten years experience, in Proc. 1st Int. Symp. Preharvest Sprouting in Cereals, Stoy, V. and Derera, N. F., Eds., *Cereal Res. Commun.,* 4, 107, 1976.
15. **George, D. W.,** High temperature seed dormancy in wheat, *Crop Sci.,* 7, 249, 1967.
16. **Kiessling, L.,** Untersuchungen über die Keimreife der Getreide, *Landwirtsch. Jahrb. Bayern,* 1, 449, 1911.
17. **Nilsson-Ehle, H.,** Zur Kenntnis der mit der Keimungsphysiologie des Weizens in Zusammenhang stehenden inneren Faktoren., *Z. Pflanzenzuecht.,* 2, 153, 1914.

18. **Gfeller, F. and Svejda, F.,** Inheritance of post-harvest seed dormancy and kernel color in spring wheat lines, *Can. J. Plant Sci.*, 40, 1, 1960.
19. **Everson, E. H. and Hart, R. B.,** Varietal variation for dormancy in mature wheat, *Q. Bull. of Mich. Agric. Exp. Stn.*, 43, 820, 1961.
20. **Freed, R. D.,** Seedcoat Color and Dormancy in Wheat, *Triticum aestivum* L., Ph.D. thesis, Michigan State University, East Lansing, 1972.
21. **Freed, R. D., Everson, E. H., Ringlund, K., and Gullord, M.,** Seedcoat color in wheat and the relationship to seed dormancy at maturity, in Proc. 1st Int. Symp. Preharvest Sprouting in Cereals, Stoy, V. and Derera, N. F., Eds., *Cereal Res. Commun.*, 4, 147, 1976.
22. **Reitan, L.,** Genetical aspects of seed dormancy in wheat related to seed coat colour in an 8 × 8 diallel cross, in Proc. 2nd Int. Symp. Preharvest Sprouting in Cereals, Gale, M. D. and Stoy, V., Eds., *Cereal Res. Commun.*, 8, 275, 1980.
23. **McEwan, J. M.,** The sprouting reaction of stocks with single genes for red grain colour derived from Hilgendorf 61 wheat, in Proc. 2nd Int. Symp. Preharvest Sprouting in Cereals, Gale, M. D. and Stoy, V., Eds., *Cereal Res. Commun.*, 8, 261, 1980.
24. **Sander, D. A.,** The Nature of Dormancy in Common Red Wheat, Ph.D. dissertation, Purdue University, West Lafayette, Ind., 1964.
25. **Piech, J., Ruszkowski, M., and Jaworska, K.,** Inheritance of the seed dormancy stage duration in winter wheat (*Triticum aestivum*) L., *Genet. Pol.*, 11, 227, 1970.
26. **Olsson, G.,** Breeding for sprouting resistance in wheat, in *Proc. 2nd Int. Winter Wheat Conference, Zagreb*, 1975, 108.
27. **Bhatt, G. M. and Derera, N. F.,** Potential use of Kenya 321-type dormancy in a wheat breeding programme aimed at evolving varieties tolerant to pre-harvest sprouting, Proc. 2nd Int. Symp. Preharvest Sprouting in Cereals, Gale, M. D. and Stoy, V., Eds., *Cereal Res. Commun.*, 8, 291, 1980.
28. **Bhatt, G. M., Ellison, F. W., and Mares, D. J.,** Inheritance studies in dormancy in three wheat crosses, in *Proc. 3rd Int. Symp. Preharvest Sprouting in Cereals*, Kruger, J. E. and LaBerge, D. E., Eds., Westview Press, Boulder, Colo., 1983, 274.
29. **DePauw, R. M. and McCaig, T. N.,** Recombining dormancy from RL4137 with white seed color, in *Proc. 3rd Int. Symp. Preharvest Sprouting in Cereals*, Kruger, J. E. and LaBerge, D. E., Eds., Westview Press, Boulder, Colo., 1983, 251.
30. **Miyamoto, T. and Everson, E. H.,** Biochemical and physiological studies of wheat seed pigmentation, *Agron. J.*, 50, 733, 1958.
31. **Stoy, V. and Sundin, K.,** Effects of growth regulating substances in cereal seed germination, Proc. 1st Int. Symp. Preharvest Sprouting in Cereals, Stoy, V. and Derera, N. F., Eds., *Cereal Res. Commun.*, 4, 157, 1976.
32. **Derera, N. F. and Bhatt, G. M.,** Germination inhibition of the bracts in relation to pre-harvest sprouting tolerance in wheat, in Proc. 2nd Int. Symp. Preharvest Sprouting in Cereals, Gale, M. D. and Stoy, V. Eds., *Cereal Res. Commun.*, 8, 199, 1980.
33. **Harvey, B. L., Rossnagel, B. G., and Muderewich, R. P.,** Sprouting resistance in barley, in *Proc. 3rd Int. Symp. Preharvest Sprouting in Cereals*, Kruger, J. E. and LaBerge, D. E., Eds., Westview Press, Boulder, Colo., 1983, 239.
34. **Gaspar, T., Wyndaele, R., Bouchet, M., and Ceulemans, E.,** Peroxidase and α-amylase activities in relation to germination of dormant and non-dormant wheat, *Physiol. Plant*, 40, 11, 1977.
35. **Noll, J. S.,** Peroxidases and their relationship to dormancy and germination in the wheat kernel, in *Proc. 3rd Int. Symp. Preharvest Sprouting in Cereals*, Kruger, J. E. and LaBerge, D. E., Eds., Westview Press, Boulder, Colo., 1983, 132.
36. **Procento, D. F. and Sapovalov, A. I.,** The comparative winter hardiness of certain varieties of winter wheat in relation to their state of dormancy, in *The Physiology of Dormancy*, Nauka, Moscow, 1968, 55.
37. **McCrae, D. G.,** Changes in the Growth-Regulating Substances of Winter Wheat during Cold Acclimation, Ph.D. thesis, Carleton University, Ottawa, 1978.
38. **Wightman, F.,** Modern chematographic methods for the identification and quantification of plant growth regulators and their application to studies of prechanges in hormonal substances in winter wheat during acclimation to cold stress conditions, in *Plant Growth Regulation and World Agriculture*, Scott, T. K., Ed., Plenum Press, New York, 1979, 327.
39. **Lalk, I. and Dorffling, K.,** Hardening, abscisic acid, proline and freezing resistance in two winter wheat varieties, *Physiol. Plant.*, 63, 287, 1985.
40. **Quarrie, S. A. and Cloutier, Y.,** Effects of cold hardening on abscisic acid levels in two wheat varieties differing in cold tolerance, *Plant Physiol.*, submitted.
41. **Allan, R. E. and Rubenthaler, G. L.,** quoted in Briggle, L. W., Pre-harvest sprouting damage in wheat in the U.S., in Proc. 2nd Int. Symp. Preharvest Sprouting in Cereals, Gale, M. D. and Stoy, V., Eds., *Cereal Res. Commun.*, 8, 245, 1980.

42. **Clarke, J. M.,** Wheat weathering damage in windrower/combine vs. direct-combine harvesting systems, in *Proc. 3rd Int. Symp. Preharvest Sprouting in Cereals,* Kruger, J. E. and LaBerge, D. E., Eds., Westview Press, Boulder, Colo., 1983, 287.
43. **Galindo, J. and S. Romero,** *Fusarium moniliforme* as the cause of pre-harvest sprouting of maize in Mexico, in *Proc. 3rd Int. Symp. Preharvest Sprouting in Cereals,* Kruger, J. E. and LaBerge, D. E., Eds., Westview Press, Boulder, Colo., 1983, 213.
44. **Dickson, J. G.,** Leaf and head blights of cereals, in *USDA Year Book of Agriculture 1953,* U.S. Department of Agriculture, Washington, D.C., 1953, 344.
45. **Doig, R. I., Done, A. A., and Rodgers, D. F.,** Pre-harvest sprouting in bread wheat (*Triticum aestivum*) as influenced by cytoplasmic male-sterility derived from *Triticum timopheevi, Euphytica,* 24, 229, 1975.
46. **Clayton, P. B. and Ellis, P. B.,** The influence of cross-pollination on pre-harvest sprouting in male-sterile line of winter wheat, in Proc. 1st Int. Symp. Preharvest Sprouting in Cereals, Stoy, V. and Derera, N. F., Eds., *Cereal Res. Commun.,* 4, 121, 1976.
47. **Jonsson, J. O.,** Pre-harvest sprouting problems in hybrid wheat breeding, Proc. 1st Int. Symp. Preharvest Sprouting in Cereals, Stoy, V. and Derera, N. F., Eds., *Cereal Res. Commun.,* 4, 115, 1976.
48. **Ellis, J. R. S. and Clayton, P. B.,** The influence of cross-pollination on pre-harvest sprouting in F_1 hybrid wheat seed, in Proc. 1st Int. Symp. Preharvest Sprouting in Cereals, Stoy, V. and Derera, N. F., Eds., *Cereal Res. Commun.,* 4, 127, 1976.
49. **King, R. W. and Richards, R. A.,** Water intake in relation to pre-harvest sprouting damage in wheat: ear characteristics, *Aust. Agric. Res.,* 35, 327, 1984.
50. **Gale, M. D. and Marshall, G. A.,** Insensitivity to gibberellin in dwarf wheats, *Ann. Bot. London,* 37, 729, 1973.
51. **Gale, M. D. and Marshall, G. A.,** The nature and genetic control of gibberellic insensitivity in dwarf wheat grain, *Heredity,* 35, 55, 1975.
52. **McMaster, G. J.,** Response to GA_3 in a-amylase synthesis of four wheat cultivars, in Proc. 1st Int. Symp. Preharvest Sprouting in Cereals, Stoy, V. and Derera, N. F., Eds., *Cereal Res. Commun.,* 4, 227, 1976.
53. **Bhatt, G. M., Derera, N. F., and McMaster, G. J.,** Utilisation of Tom Thumb source of pre-harvest sprouting tolerance in a wheat breeding programme, *Euphytica,* 26, 565, 1977.
54. **Flintham, J. E. and Gale, M. D.,** The Tom Thumb dwarfing gene *Rht3,* in wheat. I. Reduced preharvest sprouting damage to breadmaking quality, *Theor. Appl. Genet.,* 62, 121, 1982.
55. **Mares, D. J., Ellison, F. W., and Derera, N. F.,** Gibberellic acid insensitivity genes and pre-harvest sprouting damage resistance, in *Proc. 3rd Int. Symp. Preharvest Sprouting in Cereals,* Kruger, J. E. and LaBerge, D. E., Eds., Westview Press, Boulder, Colo., 1983, 22.
56. **Fick, G. N. and Qualset, C. O.,** Genetic control of endosperm amylase activity and gibberellic acid responses in standard-height and short-statured wheats, *Proc. Natl. Acad. Sci. U.S.A.,* 72, 892, 1975.
57. **Bhatt, G. M., Derera, N. F., and McMaster, G. J.,** Breeding white-grained spring wheat for low alpha-amylase synthesis and insensitivity to gibberellic acid in grain, in Proc. 1st Int. Symp. Preharvest Sprouting in Cereals, Stoy, V. and Derera, N. F., Eds., *Cereal Res. Commun.,* 4, 245, 1976.
58. **McCrate, A. J., Nielson, M. T., Paulsen, G. M., and Heyne, E. G.,** Gibberellic acid effects on germination and a-amylase activity of winter wheats, *Euphytica,* 30, 875, 1981.
59. **Henry, R. and McLean, B.,** Relative rates of sprouting, alpha-amylase production and endosperm breakdown during intact head wetting at different temperatures, in *Proc. 4th Int. Symp. Preharvest Sprouting in Cereals,* Mares, D. J., Ed., Westview Press, Boulder, Colo., 1986, 131.
60. **Ringlund, K.,** Differences in Falling Number at constant alpha-amylase activity, in *Proc. 3rd Int. Symp. Preharvest Sprouting in Cereals,* Kruger, J. E. and LaBerge, D. E., Eds., Westview Press, Boulder, Colo., 1983, 111.
61. **McCaig, T. N. and DePauw, R. M.,** Falling numbers and alpha-amylase in sawfly-resistant wheats, in *Proc. 3rd Int. Symp. Preharvest Sprouting in Cereals,* Kruger, J. E. and LaBerge, D. E., Eds., Westview Press, Boulder, Colo., 1983, 79.
62. **Cornforth, C. A. and Black, M.,** α-Amylase content of premature unsprouted wheat grains, *J. Cereal Sci.,* 3, 295, 1985.
63. **Marais, G. F. and Kruis, W. J. G.,** Pre-harvest sprouting — the South African situation for seed dormancy in barley, in *Proc. 3rd Int. Symp. Preharvest Sprouting in Cereals,* Kruger, J. E. and LaBerge, D. E., Eds., Westview Press, Boulder, Colo., 1983, 267.
64. **Gale, M. D. and Gregory, R. S.,** A rapid method for early generation selection of dwarf genotypes in wheat, *Euphytica,* 26, 733, 1977.
65. **Morris, R., Schmidt, J. W., and Johnson, V. A.,** Chromosomal location of a dwarfing gene in a Tom Thumb wheat derivative by monosomic analysis, *Crop Sci.,* 12, 247, 1972.
66. **Li, P. G. et al.,** [A preliminary study of dwarfism of wheat Ai bian No. 1], *Shanxi Agric. Sci.,* 4, 5, 1980 (in Chinese).
67. **Izumi, N., Sawada, S., and Sasakuma, T.,** A dominant gene of dwarfism located on chromosome 4D in *Triticum aestivum* cv. 'Ai-bian 1', *Wheat Inf. Serv.,* 53, 21, 1981.

68. **Worland, A. J.,** Gibberellic acid insensitive dwarfing genes in Southern European wheats, *Euphytica*, 353, 857, 1986.
69. **Gale, M. D. and Marshall, G. A.,** The chromosomal location of *Gai1* and *Rht1*, genes for gibberellin insensitivity and semi-dwarfism, in a derivative of Norin 10 wheat, *Heredity*, 37, 283, 1976.
70. **Worland, A. J.,** Gibberellic acid insensitive dwarfing genes from Saitama 27, *Euphytica*, 35, 857, 1986.
71. **McVittie, J. A. Gale, M. D., Marshall, G. A. and Westcott, B.,** The intrachromosomal mapping of the Norin 10 and Tom Thumb dwarfing genes, *Heredity*, 40, 67, 1978.
72. **Lenton, J. R., Hedden, P., and Gale, M. D.,** Gibberellin insensitivity and depletion in wheat — consequences for development, in *Hormone action in plant development — a critical appraisal*, Hoad, G. V., Lenton, J. R., Jackson, M. B., and Aitkin, R. K., Eds., Butterworths, London, 1987, 145.
73. **Gale, M. D. and Youssefian, S.,** Pleiotropic effects of the Norin 10 dwarfing genes, *Rht1* and *Rht2*, and interactions in response to chlormequat, in *Proc. 6th Int. Wheat Genet. Symp.*, Sakamoto, S., Ed., Plant Germ-Plasm Inst., Kyoto University, Kyoto, 1984, 271.
74. **Favret, E. A., Favret, G. C., and Malvarez, E. M.,** Genetic regulatory mechanisms for seedling growth in barley, in *Barley Genetics III, Proc. 3rd Int. Barley Genet. Symp.*, Gaul, H., Ed., Karl Thiemig, Munich, 1975, 37.
75. **Gustafson, P. J.,** personal communication, 1987.
76. **Phinney, B. O.,** Dwarfing genes in *Zea mays* and their relation to the gibberellins, in *Plant Growth Regulation*, Klein, R. M., Ed., Iowa State College Press, Ames, 1961, 489.
77. **Lenton, J. R. and Gale, M. D.,** Hormonal changes during cereal grain development, in *Proc. 4th Int. Symp. Preharvest Sprouting in Cereals*, Mares, D. J., Ed., Westview Press, Boulder, Colo., 1987, 253.
78. **Gale, M. D., Salter, A. M., Curtis, F. C., and Angus, W. J.,** The exploitation of the Tom Thumb dwarfing gene, *Rht3*, in F_1 hybrid wheats, in Semi-dwarf cereal mutants and their use in cross breeding. II, Proc. 1st FAO/IAEA Co-ordination Mtg., December, 1985, Rome Italy. IAEA TECDOC 455.
79. **Gale, M. D., Salter, A. M., and Angus, W. J.,** The effect of dwarfing genes on the expression of heterosis for yield in F_1 hybrid wheat, in use of induced mutations in connection with haploids and heterosis in cereals, Proc. FAO/IAEA Co-ordination Mtg., IAEA TECDOC, December 1986, Guelph, Ont., in press.
80. **Weilenmann, F.,** A selection method to test the sprouting resistance in wheat, in Proc. 1st Int. Symp. Preharvest Sprouting in Cereals, Stoy, V. and Derera, N. F., Eds., *Cereal Res. Commun.*, 4, 267, 1976.
81. **Derera, N. F., McMaster, G. J., and Balaam, L. N.,** Pre-harvest sprouting resistance and associated components in 12 wheat cultivars, in Proc. 1st Int. Symp. Preharvest Sprouting in Cereals, Stoy, V. and Derera, N. F., Eds., *Cereal Res. Commun.*, 4, 173, 1976.
82. **McIntosh, R. A.,** A Catalogue of Gene Symbols for Wheat (1983 Edition), *Proc. of 6th Int. Wheat Genet. Symp.*, Sakamoto, S., Ed., Plant Germ-Plasm Inst., Kyoto University, Kyoto, 1983, 1197.
83. **Baker, R. J.,** Inheritance of seed coat colour in eight spring wheat cultivars, *Can. J. Plant Sci.*, 61, 719, 1981.
84. **Kimber, G.,** The inheritance of red grain colour in wheat, *Z. Pflanzenzuech.*, 66, 151, 1971.
85. **Gordon, I. L.,** Heritability of grain development traits associated with sprouting damage in wheat inbreds, in Proc. 2nd Int. Symp. Preharvest Sprouting in Cereals, Gale, M. D. and Stoy, V., Eds., *Cereal Res. Commun.*, 8, 185, 1980.
86. **Belderok, B.,** Studies on dormancy in wheat, in *Proc. Int. Seed Test. Assoc.*, 26, 697, 1961.
87. **Belderok, B.,** Changes in the seed coat of wheat kernels during dormancy and after-ripening, in Proc. 1st Int. Symp. Preharvest Sprouting in Cereals, Stoy, V. and Derera, N. F., Eds., *Cereal Res. Commun.*, 4, 165, 1976.
88. **Fong, F., Koehler, D. E., and Smith, J. D.,** Genetic and chemical control of carotenogenesis and vivipary in maize seed, *Plant Physiol.*, 67(Suppl.), 142, 1981.
89. **Fong, F., Koehler, D. E., and Smith, J. D.,** Fluridone induction of vivipary during maize seed development, in *Proc. 3rd Int. Symp. Preharvest Sprouting in Cereals*, Kruger, J. E. and LaBerge, D. E., Eds., Westview Press, Boulder, Colo., 1983, 188.
90. **Robichaud, C. S., Wong, J., and Sussex, I. M.,** Control of *in vitro* growth of viviparous embryo mutants of maize by abscisic acid, *Dev. Genet.*, 1, 325, 1980.
91. **Taylor, I. B.,** ABA deficient tomato mutants, in *Developmental Mutants in Higher Plants*, Thomas, H. and Grierson, D., Eds., SEB Seminar Series, Cambridge University Press, Cambridge, 1987, 197.
92. **Ibrahim, H. A.,** Studies on the sprouting problem in soft white wheats, *Diss. Abstr. B*, 28(1), 16, 1967.
93. **Quarrie, S. A.,** Genetic variability and heritability of drought-induced abscisic acid accumulation in spring wheat, *Plant Cell Env.*, 4, 147, 1981.
94. **Neill, S. J., Horgan, R., and Watson, D. C.,** Biosynthesis of abscisic acid, in *The Biosynthesis and Metabolism of Plant Hormones*, Crozier, A. and Hillman, J. R., Eds., SEB Seminar Series, Cambridge University Press, Cambridge, 1984, 43.
95. **Koorneef, M., Jorna, M. L., Brinkhorst-van der Swan, D. L. C., and Karssen, C. M.,** The isolation of abscisic acid (ABA) deficient mutants by selection of induced revertants in non-germinating gibberellin sensitive lines of *Arabidopsis thaliana* (L.) Heynh., *Theor. Appl. Genet.*, 61, 385, 1982.

96. **Simmonds, N. W.,** Linkage to the S-locus in diploid potatoes, *Heredity*, 21, 473, 1966.
97. **Quarrie, S. A. and Lister, P. G.,** Evidence of plastid control of abscisic acid accumulation in barley (*Hordeum vulgare* L.), *Z. Pflanzenphysiol.*, 114, 205, 1984.
98. **Derera, N. F., Bhatt, G. M., and McMaster, G. J.,** On the problem of pre-harvest sprouting of wheat, *Euphytica*, 26, 299, 1977.
99. **Noll, J. S., Dyck, P. L., and Czarnecki, E.,** Expressions of RL 4137 type of dormancy in F_1 seeds of reciprocal crosses in common wheat, *Can. J. Plant Sci.*, 62, 345, 1982.
100. **Mares, D. J.,** Pre-harvest sprouting tolerance in white grained wheat, in *Proc. 4th Int. Symp. Preharvest Sprouting in Cereals*, Mares, D. J., Ed., Westview Press, Boulder, Colo., 1987, 64.
101. **Weilenmann, F.,** Plant breeding aspects of sprouting resistance and experience with the selection technique using sprouting index, in Proc. 2nd Int. Symp. Preharvest Sprouting in Cereals, Gale, M. D. and Stoy, V., Eds., *Cereal Res. Commun.*, 8, 209, 1980.
102. **Calvert, O. H., Zuber, M. S., and Neuffer, M. G.,** Vivipary in *Zea mays* induced by *Diplodia maydis*, *Phytopathology*, 59, 239, 1969.
103. **King, R. W.,** The physiology of pre-harvest sprouting — a review, in *Proc. 3rd Int. Symp. Preharvest Sprouting in Cereals*, Kruger, J. E. and LaBerge, D. E., Eds., Westview Press, Boulder, Colo., 1983, 11.
104. **Mares, D. J.,** Preservation of dormancy in freshly harvested wheat grain, *Aust. J. Agric. Res.*, 34, 33, 1983.
105. **Mares, D. J.,** Investigation of the pre-harvest sprouting damage resistance mechanisms in some Australian white wheats, in *Proc. 3rd Int. Symp. Preharvest Sprouting in Cereals*, Kruger, J. E. and LaBerge, D. E., Eds., Westview Press, Boulder, Colo., 1983, 59.
106. **Mares, D. J.,** Rate and location of production of α-amylase in relation to pre-harvest sprouting in wheat, in *Proc. 4th Int. Symp. Pre-harvest Sprouting in Cereals*, Mares, D. J., Ed., Westview Press, Boulder, Colo., 1987, 494.
107. **Zeven, A. C.,** Tom Pouce Blanc and Tom Pouce Barbu Rouge, *two Triticum aestivum* sources of very short straw, *Wheat Inf. Serv.*, 29, 8, 1969.
108. **King, R. W.,** personal communication, 1986.
109. CIMMYT Report 1966—1967, International Maize and Wheat Improvement Center (Mexico), 77.
110. **Moss, H. J.,** The pasting properties of some wheat starches free of sprout damage, in Proc. 2nd Int. Symp. Preharvest Sprouting in Cereals, Gale, M. D. and Stoy, V., Eds., *Cereal Res. Commun.*, 8, 297, 1980.
111. **Loney, D. P., Jenkins, L. D., and Meredith, P.,** The developing starch granule, *Starke*, 27, 145, 1975.
112. **Chojecki, A. J. S., Gale, M. D., and Bayliss, M. W.,** The number and sizes of starch granules in the wheat endosperm, and their association with grain weight, *Ann. Bot. London*, 58, 819, 1986.
113. **Larson, R. I. and MacDonald, M. D.,** Cytogenetics of solid stem in common wheat. III. Culm measurements and their relation to stem solidness in monosomic lines of the variety S-615, *Can. J. Bot.*, 37, 379, 1959.
114. **Sargeant, J. G. and Walker, T. S.,** Adsorption of wheat α-amylase isoenzymes to wheat starch, *Staerke*, 30, 150, 1978.
115. **Baulcombe, D. C., Huttly, A. K., Martienssen, R. N., Barker, R. F., and Jarvis, M. G.,** A novel α-amylase gene (α-Amy 3), *Mol. Gen. Genet.*, 209, 33, 1987.
116. **Hart, G. E. and Gale, M. D.,** Guidelines for nomenclature of biochemical loci in wheat and related species, *Annu. Wheat Newsl.*, 32, 19, 1986.
117. **Olered, R. and Jonsson, G.,** Electrophoretische Studie von α-amylase in Weizen, *Getreide Mehl*, 18, 95, 1968.
118. **Marchylo, B. A., LaCroix, L. J., and Kruger, J. E.,** The synthesis of alpha-amylase in specific tissues of the immature wheat kernel, in Proc. 2nd Int. Symp. Preharvest Sprouting in Cereals, Gale, M. D. and Stoy, V., Eds., *Cereal Res. Commun.*, 8, 61, 1980.
119. **Brown, A. D. H. and Jacobsen, J. B.,** Genetic basis and natural variations of α-amylase isozymes in barley, *Genet. Res. (Camb.)*, 40, 315, 1982.
120. **Sargeant, J. G.,** α-Amylase isoenzymes and starch degradation, in Proc. 2nd Int. Symp. Preharvest Sprouting in Cereals, Gale, M. D. and Stoy, V., Eds., *Cereal Res. Commun.*, 8, 77, 1980.
121. **Nishikawa, K. and Nobuhara, M.,** Genetic studies of alpha-amylase isozymes in wheat. I. Location of genes and variation in tetra- and hexaploid wheat, *Jpn. J. Genet.*, 46, 345, 1971.
122. **Nishikawa, K., Furuta, Y., Hina, Y., and Yamada, T.,** Genetic studies of alpha-amylase isozymes in wheat. IV. Genetic analyses in hexaploid wheat, *Jpn. J. Genet.*, 56, 385, 1981.
123. **Gale, M. D., Law, C. N., Chojecki, A. J., and Kempton, R. A.,** Genetic control of α-amylase production in wheat, *Theor. Appl. Genet.*, 64, 309, 1983.
124. **Baulcombe, D. C. and Buffard, D.,** Gibberellic-acid regulated expression of α-amylase and six other genes in wheat aleurone layers, *Planta*, 157, 493, 1983.
125. **Lazarus, C. M., Baulcombe, D. C., and Martienssen, R. A.,** Alpha-amylase genes of wheat are two multigene families which are differentially expressed, *Plant Mol. Biol.*, 5, 13, 1985.

126. **Ainsworth, C. C., Doherty, P., Edwards, K. G. K., Martienssen, R. A., and Gale, M. D.,** Allelic variation at α-amylase loci in hexaploid wheat, *Theor. Appl. Genet.*, 70, 400, 1985.
127. **Gale, M. D. and Ainsworth, C. C.,** The relationship between alpha-amylase species found in developing and germinating wheat grain, *Biochem. Genet.*, 22, 1031, 1984.
128. **Gale, M. D. and Baulcombe, D. C.,** unpublished observation, 1986.
129. **Daussant, J. and Renard, H. A.,** Development of different α-amylase isozymes, having high and low isoelectric points during early stages of kernel development in wheat, *J. Cereal Sci.*, 5, 13, 1987.
130. **Ainsworth, C. C., Miller, T. E., and Gale, M. D.,** α-Amylase and β-amylase homoeoloci in species related to wheat, *Genet. Res.*, 49, 93, 1987.
131. **Kneen, E. and Sandstedt, R. M.,** An amylase inhibitor from certain cereals, *J. Am. Chem. Soc.*, 65, 1247, 1943.
132. **Warchalewski, J. R.,** Preliminary investigation on purification of native alpha-amylase inhibitors from durum wheat, *Bull. Acad. Pol. Sci. Biol.*, 24, 559, 1976.
133. **Munday, J., Hejgaard, J., and Svendsen, I.,** Characterisation of a bifunctional wheat inhibitor of endogenous α-amylase and subtilisin, *FEBS Lett.*, 167, 210, 1984.
134. **Weselake, R. J., Macgregor, A. W., and Hill, R. D.,** An endogenous α-amylase inhibitor in barley kernel, *Plant Physiol.*, 72, 809, 1983.
135. **Munday, J., Svendsen, J., and Hejgaard, J.,** Barley α-amylase/subtilisin inhibitor. I. Isolation and characterisation, *Carlsberg Res. Commun.*, 48, 81, 1983.
136. **Yoshikawa, M., Iwasaki, T., Fujii, M., and Oogaki, M.,** Isolation and some properties of a subtilis in inhibitor from barley, *J. Biochem. Tokyo*, 79, 765, 1976.
137. **Cubadda, R.,** Chromosomal location of genes controlling synthesis of some soluble proteins in *T. durum* and *T. aestivum*, *Genetics and Breeding of Durum Wheat*, Biblioteca della Facolta di Agraria, Mugnozza, G. T., Scarascia, Ed., Bari, 1973, 653.
138. **Aragoncillo, C., Rodriquez Loperena, M. A., Carbonero, P., and Garcia-Olmedo, F.,** Chromosomal control of non-gliadin proteins from the 70% ethanol extracts of wheat endosperm, *Theor. Appl. Genet.*, 45, 322, 1975.
139. **Konarev, A. V.,** [Identification of inhibitors of own and alien α-amylase in proteins of wheat], in *Peculiarities of Biology and Utilisation of Species and Varieties of Cultivated Plants in Breeding*, Bull. 1, Vavilov Institute of Plant Industry, Fasc. 118, ISSN 0202, 1982 (in Russian).
140. **Sanchez-Monge, R., Barber, D., Mendez, E., Garcia-Olmedo, F., and Salcedo, G.,** Genes encoding α-amylase inhibitors are located on the short arms of chromosomes 3B, 3D and 6D of wheat *(T. aestivum)*, *Theor. Appl. Genet.*, 72, 108, 1986.
141. **Bedetti, C., Bozzini, A., Silano, V., and Vittozzi, L.,** Amylase protein inhibitors and the role of *Aegilops* species in polyploid wheat speciation, *Biochem. Biophys. Acta*, 362, 299, 1974.
142. **Piazzi, S. E., Riparbelli, G., Sordi, S., Cantagalli, P., Pocchiari, F., and Silano, V.,** Immunochemical characterisation of specific albumins of bread wheat, *Cereal Chem.*, 49, 72, 1972.
143. **Hejgaard, J., Bjorn, S. E., and Nielsen, G.,** Localisation to chromosomes of structural genes for the major protease inhibitors of barley grains, *Theor. Appl. Genet.*, 68, 127, 1984.
144. **Peruanskii, Y. V., Gabsattarova, B. S., and Peruanskaya, O. N.,** [Studies of the composition and activity of wheat α-amylase inhibitors in relation to growth conditions], *Fiziol. Biokhim. Kult. Rast.*, 12, 358, 1980 (in Russian).
145. **Gale, M. D. and Salter, A. M.,** unpublished result.
146. **Olugbemi, L. B., Austin, R. B., and Bingham, J.,** Effects of awns on the photosynthesis and yield of wheat, *Triticum aestivum*, *Ann. Appl. Biol.*, 84, 241, 1976.
147. **Silano, V.,** α-Amylase inhibitors, in *Enzymes and Their Role in Cereal Technology*, Kruger, J. E., Lineback, D., Stauffer, C. E., Eds., American Association of Cereal Chemists, St. Paul, 1987, 141.

Chapter 6

BREEDING FOR PREHARVEST SPROUTING TOLERANCE

N. F. Derera

TABLE OF CONTENTS

I. INTRODUCTION

A review of all approaches for combating preharvest sprouting damage suggests that the most feasible approach is to incorporate genes for preharvest sprouting resistance in our cereal cultivars. Breeding for preharvest sprouting resistance in cereals was initiated early this century. However, these sporadic efforts resulted in only some degree of success. During the 1940s this problem was neglected and the emphasis shifted to yield improvement and disease resistance. Increased interest in preharvest sprouting research and, consequently, breeding for sprouting resistance occurred in the early 1970s. Well organized systematic breeding programs were initiated in Canada, Sweden, England, Norway, Finland, Poland, West Germany, Australia, and Switzerland.

Breeding for preharvest sprouting resistance is an involved task. No general solution can be given. Each region needs, according to its growing practice and environment, a specific combination of resistance factors; single resistance factors will not give adequate protection in every season. For example, in many wheat growing districts where winter wheat is grown, long-lasting dormancy genes are unacceptable due to seasonal time constraints. In these situations an alternative strategy such as slow water uptake is needed. It is also noted that the strongest dormancy can be broken quickly if the crop is exposed to intermittent rains and drying out periods at high temperatures after translocation has ceased.[1] Factors for sprouting resistance other than dormancy need to be identified.

II. BREEDING OBJECTIVES

One of the requirements of a successful breeding program is that its objectives be clearly determined. After these objectives have been set, an appropriate strategy can be defined. The majority of researchers into preharvest sprouting resistance continue to base their breeding effort on dormancy factors. Dormancy genes appear to be easier to handle than the other factors as recently our knowledge on dormancy and its inheritance have widened considerably.

Derera advocates that other factors, including the rate of water uptake in the caryopsis, starch resistance to amylases causing degradation, reduced rate of amylase production, and the physical characteristics of the spike, should all be considered when setting out the aims of a cereal improvement program where preharvest sprouting resistance is one of the breeding objectives.[1]

Early or late ripening traits may be utilized to avoid sprouting damage. Ringlund believes that in Norway the early maturing barley cultivars may escape the preharvest rains more than the later ripening ones.[2] However, he also warns that early cultivars may have less dormancy.

In the regions where wind storms are frequent around harvest time, lodging resistance is an important factor to minimize the effect of the rain during the preharvest period.

Spikes with awns take up slightly more moisture than awnless spikes and do so faster. The so-called club spike also absorbs more water than the tapering one. Moreover the angle of the spike should be considered. Nodding of wheat spikes rarely exceeds 90° while barley has a nodding angle of more than 120° from the vertical. A horizontal wheat spike takes up more water than does a vertical spike. The 120° nodding of barley may be able to shed water and so could slow down water uptake.[4] Closed glumes versus open "gaping" ones can also hinder the access of water to the grain.

Perhaps in the not-too-distant future a germless grain can be produced. At present, this may sound far-fetched; however germless grains are found routinely in many cereals, e.g., the New York state wheat cultivar Yorkstar produces approximately 10% germless grains. The embryoless grain is caused by the single fertilization of the polar nuclei. The germless grain cannot sprout, so its use would be ideal. This highly innovative approach is useful in a hybrid seed production system. Marshall recommends three ways to produce cultivars or hybrids:

1. Using dominant complementary genes which produce embryo lethality in the progeny
2. Using a recessive embryo lethal closely linked to a gametophytic factor or gamete eliminator
3. Applying embryo lethals whose expression is environmentally (physically or chemically) dependent. (This system may also be used with pure line cultivars.)

The germless grain would be useful where the germinative capacity of the grain is not important: wheat, rye, triticale, rice; however, in the case of malting grains (e.g., barley) it would be useless.[5]

After a review of the breeding objectives a number of representative ideotypes that appear suitable for the two major crops, wheat and barley, can be demonstrated. These ideotypes give only a broad sense of direction and each breeding program should create its own theoretical ideotypes which best take into account their general breeding program and environmental factors.

Wheat

Plant	Erect, lodging resistant
Spike	Erect, tapering, awnless
Glumes	Closed
Earliness	A maturity range that escapes the rainy season
Kernel	Red pericarp (if trade permits), dormant, slow water uptake; low rate of amylase synthesis, starch resistant to amylase-caused degradation, hybrid cultivar to produce germless grain

Barley

Plant	Erect, lodging resistant
Spike	Nodding, more than 120° from the vertical, if possible, awnless or with rudimentary short awns
Earliness	To be adjusted to escape a rainy harvest
Kernel	Short positive dormancy of not more than 2 to 3 weeks

The classical cereal improvement project is still based on various methods of crossing and selection. To date, these methods of improvement have proven to be both economical and effective.

Other sophisticated approaches to create, or incorporate, variability such as mutation, breeding somatic recombination, and genetic engineering, are still in an early stage. No attempts have so far been reported.

Without doubt, genetic engineering opens the possibility of transferring useful genes, including those for preharvest sprouting resistance, from foreign genera to cultivated cereals. It is envisaged that a complete control of preharvest sprouting may be achieved with the help of controlled somatic hybridization and/or using recombinant DNA techniques on sprouting resistant grasses and cultivated cereals. Seemingly the routine application of genetic engineering in cereal improvement is very avant-garde; nevertheless, if the present progress is maintained, we may see cereals with foreign genes that are easy to identify and clone within the next 10 years. In the meantime, the cereal breeder must use the conventional methods and, after examining the breeding aims and ideotypes, a review of the recommended sources or potential parents merits attention. In Table 1 are listed the sources of sprouting resistance recommended and/or used since 1973 in wheat, barley, and rye programs.

III. BREEDING STRATEGIES

The breeding strategy varies from breeder to breeder, and each breeder usually modifies his/her breeding strategy in accordance with the program goal(s). Many cereals breeders approach preharvest sprouting resistance as a complex trait involving dormancy, starch sensitivity, α-amylase synthesis, response to gibberellic acid, physical characteristics of the spike, etc. Obviously large number of genes are involved and, in most cases, individual genes have not been located to their respective chromosome(s) and, in consequence, the breeder is well advised to treat preharvest sprouting resistance, RCP, as a quantitative trait. However, in the not too distant past, breeders always associated preharvest sprouting resistance with dormancy. One major objective was to determine how many genes control dormancy. Breeding, based on pedigree selection, testing germination energy and power at a given time after harvest ripeness of F_2- and F_3-derived lines, was used to ascertain the degree of dormancy present.[27]

A. Red Dormant Cultivars

The Swedish Seed Association program in the early 1970s followed the pedigree and backcross breeding methods. Their selection strategy was based on the theory that dormancy is a dominant trifactorial trait. Nevertheless, they suggested that "many other genes must have influence upon the damage by sprouting". The Swalöf program also experienced transgressional segregation of dormancy. Their testing for dormancy went well beyond just having a simple germination test. A number of spikes of a single plant progeny line were treated in a rainfall simulator and then the α-amylase activity was assayed with a colorimetric method. When larger samples became available in later generations the Hagberg falling number test was used. This program recognized 18 potential winter and spring wheat parent materials and produced a number of highly resistant strains using either the simple or the backcrossing techniques. At the time, the successful crosses were Kron/Eka, Ibrido/Dankowska, Frontana /2*SV01312, Breustet 6825-47/3*, Stube/3* Prince, and Frontana/3* Prince.[8]

The Research Station of Agriculture Canada, Winnipeg, Manitoba conducts a successful hard red spring wheat program and has produced an experimental line, RL4137, and a cultivar, Columbus, which can be considered as the two outstanding, highly resistant modern cultivars to sprouting damage. These two cultivars now provide the basis for many wheat improvement programs where preharvest sprouting resistance has to be considered. The importance of the accomplishment of the Winnipeg group justifies a detailed review of their strategies.

Basically, the Winnipeg group's work is based nearly wholly on dormancy, following the doctrine that dormancy is closely associated with genes for red grain color. It is assumed that where 2 or 3 loci are acting, an additive effect is expected. It is also presumed that allelic differences in the loci of red pericarp color could be causing the differences in the variation in the extent of dormancy of cultivars with equal numbers of genes for redness.[7] Another Canadian group from Swift Current suggests that RL4137 may have two or more genes controlling seed dormancy; one gene is linked with the red pericarp while the other one or more are not associated with the seed color.[28] The Winnipeg group is also aware of the possible influence of the differential water uptake of grain and the spike.

In breeding for preharvest sprouting resistance, both the pedigree and backcross techniques are practiced at the Winnipeg Research Station. The program to establish a sprouting-resistant cultivar with similar agronomic characteristics to 'Neepawa' started in 1969. 'Neepawa' the

Table 1
POTENTIAL SOURCES OF SPROUTING RESISTANCE

Species/Variety/Cultivar	Resistance factor	Ref.
Wheat		
RL 4137	Dormancy	6
Columbus	Dormancy	7
Kolibri	Dormancy	8
Breustedt 6825-47	Dormancy	8
Frontana	Dormancy and low α-amylase synthesis	8
Strube 4926	Dormancy	8
Breustedt 460/42	Dormancy	8
Park	Dormancy	6
Takahe	Dormancy	9
Kleiber	Dormancy	10
Bezostaja 1	Dormancy	11
Leader	Dormancy	12
7722	Dormancy	12
Kenya 321.BT.1.B.1.	Dormancy	6
Losprout	Dormancy	13
Clark's Cream	Low α-amylase synthesis	14
Eagle	High fiber, endosperm tolerance to α-amylase	15
Salamanca	Low α-amylase synthesis	16
Gwebi	Low α-amylase synthesis	16
AUS 109 — Chile 59	RCP	17
AUS 320 — Hellas	RCP and dormancy	17
AUS 471 — Lerma 52	RCP	17
AUS 633 — Mexican 852	RCP	17
AUS 758 — Pakistan C22F	RCP	17
AUS 1293 — South African 1166	RCP	17
AUS 1305 — Spoetnik	RCP	17
AUS 1408 — from Transvaal	RCP, Dormancy	17
AUS 1490 — unnamed	RCP, Dormancy	17
AUS 5820 — Kenya 59	RCP	17
AUS 7122 — Bihar 124	RCP	17
AUS 13054 — Egypt 1167 — 109	RCP	17
AUS 16115 — Saberbeg	RCP	17
AUS 16116 — Homera	RCP	17
Kite	Slow water uptake; low α-amylase synthesis	18
Shortim	Low α-amylase synthesis	18
SUN 44E (Suneca)	Low α-amylase synthesis	18
Heron	Slow water uptake by the spike and kernel	19
Warigal	Slow water uptake by the spike and kernel	19
Tom Thumb	Gibberellin insensitivity *Gai3*	20
Minister Dwarf	Gibberellin insensitivity *Gai3*	21
Tordo	Gibberellin insensitivity *Gai3*	10
Barley		
WM 143-2-3b	Dormancy	22
Agneta	Dormancy	23
Hiproly	Chymotrypsin inhibitor I and II	24
Rye		
Otello	Low α-amylase synthesis	25
Secale silvestre	Dormancy	26

recurrent parent, was used as a female parent during the backcrosses. Sprouting tests of progenies which were used as male parents for backcrossing were initiated after the second backcross and were continued until after the fourth backcross. After the second backcross, approximately 25 plants, which represented five crosses, were cultivated at the following four stages of backcrossing. The presence of dormancy in the seeds of F_1 backcross plants was ascertained by simple germination test in petri dishes at 20°C and was compared to 'Neepawa' and 'RL4137'.[7] This method proved to be very successful; nevertheless it was felt that instead of a germination test, a simple micro-amylase test would have been quicker, simpler, and perhaps equally, if not more, effective. The backcrossing was completed in 2 years and the most resistant F_2 progeny of the most successful backcross were propagated. Selected F_3 lines were then sown in a field nursery. Subjective estimation of sprouting resistance was conducted on the F_3, F_4, and F_5 lines and only the most resistant lines were kept for further study. Seed increase commenced on the F_5 lines. In yield tests during the first year, considerable genetic variation was experienced for agronomic and quality traits. It was also surmised that several genes for dormancy, or more precisely for preharvest sprouting resistance, were involved in the transfer from 'RL4137'. Eventually this program resulted in the naming and release of the commercial cv. Columbus.

Similar projects were initiated to improve preharvest sprouting resistance of the cultivars Glenlea, Chester, Benito, and Katepwa. In the case of Glenlea, Chester, and Katepwa, RL4137 was used as the source of resistance. For Benito, Columbus and Takahe were used as donor sources. The Chester backcrossing program was not successful and the breeders admit that they would have been more successful testing for α-amylase activity (and hence preharvest sprouting damage) rather than dormancy.[7]

B. White Tolerant Cultivars

The program of the Winnipeg group represents a classical approach when dormancy, together with red pericarp, is an acceptable breeding goal. To produce a wheat cultivar with white kernel and with appropriate resistance to preharvest sprouting is a task which was considered inconceivable 20 years ago. In 1973, the author of this chapter strongly recommended that the Australian wheat industry should change from the white to red wheat in the regions where preharvest sprouting damage was causing considerable losses, e.g., the Australian prime hard wheat production area (northern New South Wales and southern Queensland) where the chances of receiving rain during the harvest period are high.[29] The high temperatures during the kernel filling period also appear to reduce the effect of dormancy in some cultivars. A program to find alternative source(s) of resistance in white wheats was initiated by the University of Sydney Narrabri group in 1965. However, owing to paucity of finance, this was a "low key" program. After 1973, the search for suitable resistance factors was intensified, not only at Narrabri, but in every region growing white wheats throughout the world.

The starting point for the Narrabri program was the establishment of standard screening methods followed by attempts to find sources of resistance other than dormancy. Numerous plausible and sometimes less plausible hypotheses were investigated.[10,30-31] Initially, breeding for resistance to preharvest sprouting involved the use of predicative resistance factors such as GA insensitivity genes, germination inhibitors in the bracts, low water absorbtion, low rate of amylase synthesis, and starch resistance to α-amylase induced degradation. During the selection program, each factor was not independently tested; the only assessment was based on the actual damage. Tests included germination of unthreshed kernels within intact spikes in a rain simulator chamber followed by an α-amylase assay where practicable. The performance of advanced lines was tested with the aid of the Hagberg falling number test.

In summary, the incorporation of the gibberellin insensitivity gene *Gai3* was not successful. It seems the *Gai3* and *Rht3* genes had pleiotropic effects, and therefore plants with these genes were too short for commercial production. Nevertheless, some cultivars with *Rht2* and/or a combination of *Rht1* and *Rht2* have slower rates of α-amylase synthesis; therefore their use as a criterion in breeding for RCP is justified.[16]

The Narrabri group eventually chose not to manipulate the alleged germination inhibitors in bracts because these were highly influenced by environment. It was also assumed that the early result may have been due to an artifact(s) in the testing technique.[32] However, Harvey et al.[22] found a satisfactory proof of a germination inhibitory effect of the barley hull.

To utilize a dormancy factor present in the white-grained spring wheat 'Kenya 321 BT.1.B.1'(hereafter referred to as "Kenya 321 sib") the Narrabri group crossed it with the cultivar Shortim, which has a combination of the *Rht1* and *Rht2* dwarfing genes plus a relatively slow rate of α-amylase synthesis. The F_1 and F_2 generations were harvested in bulk. The F_3 population was grown parallel with the parents Kenya 321 sib and 'Shortim'. At anthesis, 266 random F_3 plants and 15 random plants from each of the parents were selected. The data on harvest ripeness and plant height was recorded for these plants and each plant was harvested and individually thrashed. Three samples of 50 seeds each were made ready for germination tests 10, 20, and 30 days after harvest ripeness. The germination tests were conducted in the dark and data recorded after 96 hr. The frequency distribution for the germination rate of F_4 embryos at 10, 20, and 30 days after harvest ripeness is shown in Table 2.

Considering the large gap between the dormant vs. the rest of the segregates, the differences in dormancy could be accepted as qualitative. This particular study suggests that the dormancy factor in Kenya 321 sib is recessive and that this type of dormancy gives only very brief protection against preharvest sprouting. Nevertheless, the 6% short dormant types recovered among the randomly selected F_3 plants and the nonsignificant correlation between plant height and this particular dormancy offer promise in the use of Kenya 321 sib in white wheat breeding programs. Notwithstanding that Kenya 321 sib seems to be a useful source of dormancy among white-kernelled wheats, one notes that its agronomic performance and kernel quality are inferior and hence its use must be restricted for backcrossing programs as the donor, nonrecurrent parent.

The paucity of sources of sprouting-resistant factors in white kernelled wheat has forced the white wheat-breeding organizations to search for new, genetic traits which could be used in the development of sprouting-resistant white-kernelled wheat cultivars. Kenya 321 sib is one such source; yet its dormancy is short and, under high temperature ripening, it is rendered useless. The shortcomings of Kenya 321 sib and the failure to utilize the gibberellin insensitivity led the Narrabri group to intensify their search for alternative sources of sprouting resistance. Using established technologies, more than 3000 white-kernelled cultivars were screened for sprouting resistance over a period of 3 years.[31,17-18] The end product of this program is 14 white-kernelled wheat cultivars with considerable sprouting tolerance that may be usable as parents in breeding sprouting-resistant cultivars. These cultivars are included in Table 1. The exact mechanism and genetics of these newly found sources of resistance will no doubt be evaluated so that their full potential can be utilized.

Mares also studied the sprouting tolerance of *Triticum* species, triticale and barley.[34] Among the diploid and tetraploid *Triticum* species, he found 6 and 26 entries, respectively, which could be used as strong sources of resistance. Among the hexaploid accessions, 11 deserve attention. Particularly interesting are the observations on *T. sphaerococcum* varieties where, besides dormancy, restricted rate of water movement to the kernel was detected in the ripe spike. Nevertheless, some of these results are based on 1-year experiments and need further verification.

For commercial cultivation, the Narrabri group have recently released two cultivars, Suneca and Sunelg, with tolerance to preharvest sprouting; however this tolerance is inferior to that of the newly discovered 14 white-kernelled, resistant cultivars as demonstrated in Table 3.

The wheat breeders of Agriculture Canada, Swift Current, Saskatchewan endeavored to transfer dormancy and the high degree of preharvest sprouting resistance from RL4137, red

Table 2
FREQUENCY DISTRIBUTION OF F_4s OF THE WHEAT CROSS 'SHORTIM' × 'KENYA 321' FOR GERMINATION PERCENTAGE AND ESTIMATES OF STATISTICAL PARAMETERS[a]

Percentage germination (class range)	Frequency in tests conducted at 10 days after harvest ripeness	20 days after harvest ripeness	30 days after harvest ripeness
11—15	1	0	0
16—20	1	0	0
21—25	1	0	0
26—30	0	0	0
31—35	2	0	0
36—40	3	0	0
41—45	3	1	0
46—50	5	0	0
51—55	2	0	0
56—60	11	1	0
61—65	11	2	0
66—70	15	7	1
71—75	13	8	0
76—80	19	10	3
81—85	27	3	0
86—90	46	8	4
91—95	42	15	7
96—100	64	13	4

Statistical parameters

No. of F_4s	266	68	19
Range of variation (%)	14—100	40—100	64—98
Mean (%)	82.76	83.38	87.89
Variance	274.15	153.44	93.67
C.V. %	20.00	14.86	11.01

[a] The vertical solid and broken lines indicate the range of variation in the two parents, 'Shortim' and 'Kenya 321' sib, respectively.

From Bhatt, G. M. and Derera, N. F., *Cereal Res. Commun.*, 8, 291, 1980. With permission.

spring wheat, to a white wheat. They used 7722 as the white-seeded spring parent. In the selection process a modified pedigree selection was applied. From this cross 'Losprout', a white-seeded spring wheat germplasm line with a reasonable preharvest sprouting resistance, was released in 1984.[13] This work illustrates, in a practical way, that certain factors of dormancy are not linked or pleiotropic with red pericarp and can hence be transferred to white-seeded wheat cultivars.

DePauw and McCaig gave some details of the techniques used for screening of sprouting resistance and kernel color and how to ascertain recovery of sprouting resistance in the newly developed genotype.[36] The first step was to verify which technique was the most practical to use in their investigation. There were 23 genotypes with varying degrees of redness of the kernel, maturity, sprouting resistance, and α-amylase activity used to test the methods. The different genotypes were grown in the field in a replicated manner and when the kernel, during its ripening process, attained 18% moisture content, the sampling was

Table 3
RELATIVE PREHARVEST SPROUTING TOLERANCE OF SOME WHITE WHEATS COMPARED WITH THE TOLERANT CANADIAN RED WHEAT RL 4137

Increasing tolerance

Strain

Standards

RL 4137

Kenya 321

Kite, Spica

Cook

Songlen

SUN 56A

Recently developed cultivars

Suneca

Sunleg

Recent lines and cultivars 1981—84

Newly found sources of tolerance

AUS 1408, AUS 1490

Saberbeg, Hellas

South African 1166, Homera, Chile 59

Kenya 59

Mexican 852, Pakistan C228

Egypt 1167-109

Bihar 124, Spoetnik, Lerma 52

Data from Reference 17.

commenced. At this time 10 random spikes were taken per plot of each genotype and stored at −23°C. This sampling was repeated 7 days later. Spikes from both samplings were placed in a controlled environment cabinet and over a 3-hr period were subjected to 135 mm of simulated rain. A further 15 mm of mist was added each successive day. The spikes were kept in the cabinet for 7 days under nearly 100% relative humidity at 20°C. On the 7th day the spikes with visible sprouts were counted. The samples were then dried at 50°C and hand-thrashed. Germination percentage on the dried sample was also ascertained. α-Amylase activity was determined using the Heygaard and Gibbons gel-diffusion technique.[35] Untreated kernels from the field were also subjected to α-amylase assay using both the gel-diffusion and Hagberg falling number techniques.

After testing the 23 genotypes by the techniques described in the foregoing paragraph, it was resolved that the observable signs of sprouting in unthreshed spikes correlated well with the percentage of germinated kernels in the spike and with α-amylase activity. It was evident that observing sprouted spikes needed considerably less time (1/100th) than counting germination percentage or assaying α-amylase activity. These investigators hence recommend the use of 10 spikes per line in a rain simulator to observe visible sprouting; this allows a rapid differentiation between sprouting resistant or susceptible progenies, lines, cultivars, etc.

The preceding technique, the observation of visible sprouting, was used to recover white sprouting-resistant segregates from populations where the sprouting resistant donor parents were RL4137 (red, dormant) and K321 (white, dormant). DePauw and McCaig found that in all three populations involving RL4137, some white-kernelled sprouting-resistant selections had better sprouting resistance than some of the red-kernelled checks. They found the best level of sprouting resistance in progenies selected from a population which originated from a backcross where RL4137 was used as a recurrent parent twice. It was suggested that the sprouting resistance factor which is independent of kernel color in RL4137 might be controlled by more than one pair of genes.

When K321 was used as a donor parent, DePauw and McCaig were able to recover a low level and "short lived" sprouting resistance. It was therefore concluded that RL4137 was a more relevant parent than K321.

The same authors also recommended a single sampling time, a few days after harvest ripeness, when only a small seed sample was available as a result of early generation selection. However, the use of two samplings is strongly advised in subsequent generations.[36]

DePauw and McCaig, in their previously quoted classic work, consistently abbreviate sprouting resistance as "SR". This can be confusing, because for decades the wheat rust researchers and breeders designated stem rust as SR. The author of this chapter has strongly recommended the use of "RCP" as the code for unspecified sprouting resistance.[1,37]

C. Mass Selection

Pedigree selection has always been an effective tool for plant improvement. Nevertheless, this technique is both time consuming and labor intensive. To reduce cost and increase the frequency of desirable genotypes in a segregating population, before single plants are selected, mass selection techniques can be practiced. During the F_2 to F_4 generations, the segregating bulk populations are exposed to appropriate selection pressures. The unwanted genotypes are largely eliminated and hence the ratio of desirable genotypes should increase so that the chance of selecting suitable recombinants in the F_5 will be significantly augmented.[38-39] Selection *en masse* for preharvest sprouting resistance (RCP) can be performed on the detached spikes or on the threshed kernels.

Svensson recommends that approximately 1000 main spikes from each population be removed at "binder ripeness" and the detached spikes placed in a rain simulator chamber where the relative humidity is kept near 100% with the help of a defensor and the temperature at 15°C (± 1°C).[40] More recently, McMaster and Derera recommended an initial use of 50 mm of simulated rain at 18 to 20°C in the chamber.[31] The technological aspects of these methods are discussed in Chapter 7. In the F_2 generation, the spikes are removed from the chamber when approximately 10% of the spikes show no visible sign of germination. The seemingly ungerminated spikes are then dried and threshed. Progenies of this material are subjected to the same process; however, in the second and third stage of mass selection a higher ratio of ungerminated spikes should be retained. The ratio should be set according to the selection pressure required.[40] Particular care should be exercised when the spikes are selected. Each ripeness group should be tested separately; otherwise the population may shift toward the later maturing types.

The degree of ripeness in mass selection and in general screening operations is most important. The various ripeness stages quoted by a number of authors are confusing. They mention binder ripeness, harvest ripeness, and embryo, morphological, and physiological ripeness. Most of these are very difficult to specify. Binder and harvest ripeness are interpreted differently in each cereal-growing region. In Canada, it is defined by a kernel moisture content of around 18%, while in Australia the designated figure is under 13% moisture content.

"Number of days after heading or anthesis" is used by several authors to specify the developmental stage when certain tests or observations are made.[41-42] This would be applicable if the environment at all sites and in all years and experiments, were strictly constant. Many environments, however, are variable, and after specified periods following heading or anthesis may show considerable variation in growth, development, and/or dehydration stage of the spike or kernel. It is, therefore, imperative that the screening of spikes or kernels be conducted at a constant stage of ripeness to avoid biased selection results, especially unwanted shifts towards lateness. A number of researchers found that the so-called eosin test is a useful starting point to specify the degree of ripeness more accurately. The principle of this method is based on the assessment of the stage when water movement to the spike has ceased; this also means the cessation of dry matter translocation to the kernel from the rest of the plant.[43] This stage may be considered as morphological ripeness. This test is further discussed in Chapter 7.

The following scores are recommended when the eosin test is conducted:

0 = No eosin coloration can be detected in the spike anywhere, although eosin may have moved in the stem to the base of the spike without entering into any of the bottom part of the rachis or glumes, etc.
1 = The lower part of the rachis is colored by the eosin but the spikelets are free of eosin.
2 = The rachis shows strong uptake of eosin; the glumes on the bottom spikelets are infused with eosin.
3 = The whole spike is colored.

For a simple quantitative expression of ripening time, it is advisable to give a value of one for the earliest genotype and add a further one for each day the genotypes take to reach eosin ripeness.[44]

With the help of the foregoing method, a fairly accurate estimation can be made of binder- or harvest-ripeness and the maturity grouping can be established with confidence.

A further mass selection method is that of germinating threshed kernels. This method is based mainly on the likely presence of dormancy factors.[45-46] This technique is simple. Hand threshed kernels, collected from randomly selected F_2, etc., spikes, at a certain period after harvest, are germinated in sand or filter paper for 3 to 4 days, depending on the degree of selection pressure required. After the preset germination time, the ungerminated kernels are removed, gently dried, and used to sow the next generation. This procedure may be repeated in the F_3 and F_4 generations prior to single plant selection. The disadvantage of this technique is that the kernels of the early and late ripening genotypes are mixed, and a shift towards later ripening types can be expected. Also, with the screening of germinating kernels, the emphasis is on dormancy, and other possible factors contributed by the intact spike are neglected.

After elimination of the highly sprouting-susceptible components of a population, single plant or single spike selection can be commenced. Single plant selection is strongly recommended, because this gives one the opportunity to perform wider tests. This is particularly true of space planted populations where single plants provide several spikes, allowing one to test plant resistance at two or three intervals. A single plant may also provide enough kernels to conduct other tests for quality, disease resistance, etc.

A representative sample spike of a single plant is placed in a rain simulator and exposed to the predetermined amount and intensity of rain and kept at near 100% RH till a portion of the spike shows visible signs of sprouting. Once again, the extent of treatment is based on the selection pressure applied during the previous mass selection stages. Spikes demonstrating sprouting resistance (and/or residue seed) provide the embryos for the next generation. The progenies of the single plant, represented by two or three short rows in the field, can be subjected to more detailed tests, and both the extent and degree of sprouting resistance can be quantified.

Besides assessing sprouting resistance at the same stage of ripeness, it is also desired to have an index expressing quantitatively the resistance at a given ripeness stage. Workers at the Swiss Federal Research Station at Zürich-Reckenholz developed an index which they and other breeders have used during the past 10 years.[47-49] The index is based on the Hagberg falling number test. Each sample, harvested on certain days after eosin ripeness, is subdivided into two: one half is threshed and the ground kernels are subjected to the falling number test without treatment; the other half is exposed to 3 days of moisture chamber (rainfall simulator) treatment and then dried, ground, and subjected to the falling number test. These tests have been called "falling number dry" and "falling number wet", respectively. These two tests, and the information on morphological ripeness, led Weilenmann to the development of an empirical formula which expresses the degree of sprouting resistance on a scale of values from 1 to 9, 9 indicating the total absence of resistance, and 1 indicating a high degree of resistance. The formula is

$$\text{Sprouting index (SI)} = \sqrt{[R - \frac{\text{Falling number dry}}{60}]\ [\frac{540}{\text{Falling number wet}}]}$$

R indicates the number of days from eosin or morphological ripeness to the day when the test was made. It is recommended that breeding lines be tested exactly 10 days after "eosin ripeness". For a more detailed study, e.g., on new cultivars, when the intensity and extent of sprouting resistance needs to be quantified and expressed with a single value, a number of such indices may be calculated, for example, for 10, 15, 20, 25, and 30 days after eosin ripeness. Each sampling-date result can be marked on a graph which will show a curve and the area above the curve is calculated (see Figure 1). A large area indicates a high degree of long-lasting sprouting resistance. Weilenmann also uses ecoanalytical examinations to analyze the data resulting from different environmental conditions. The so-called eco-values are particularly useful when the desired data are from several years of trials.[48]

Strand advocates a dormancy index (DI) which is calculated as the mean percentage of dormant kernels obtained at two germinating temperatures, in moist sand at 12 and 20°C for 10 days germination time. He gives double weight to the results at 12°C. The germination tests are conducted 10, 20, and 30 days past yellow ripeness. The visual assessment of "yellow ripeness", however, is subjective and could be a cause of error. Nevertheless, this method provides an objective assessment of dormancy by taking into consideration the cultivar × germination temperature interactions.[50]

The quantitative screening techniques discussed in the foregoing paragraphs are indirectly supported by the intensive studies of Gordon. He concludes in a number of his studies that screening of lines for sprouting resistance would be more reliable if it were based directly on sprouting attributes and that monitoring of ripeness is beneficial to obtaining objective results.[51]

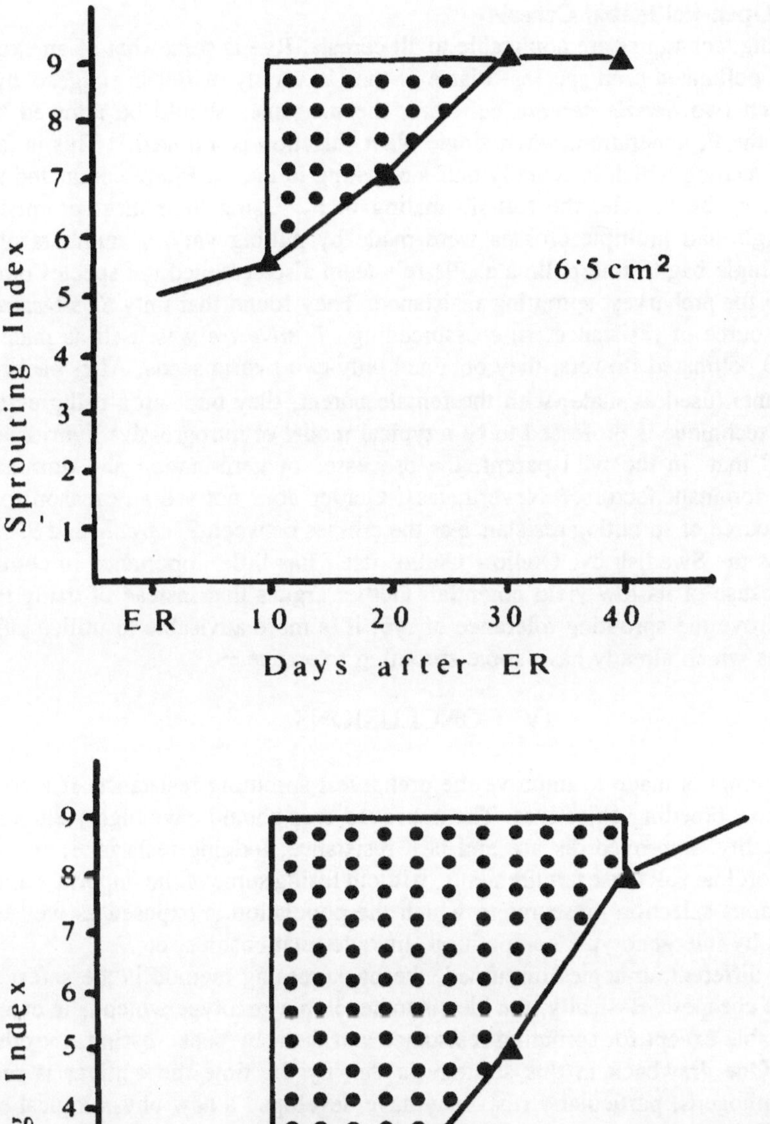

FIGURE 1. Sprouting curves of cultivars with low (top) and high degree and lasting tolerance (bottom) to preharvest sprouting. (Modified after Weilenmann, F., Ph.D. Diss. Nr. 5712, ETH Zurich, 1976. With permission.)

D. Breeding Open-Pollinated Cereals

Most breeding techniques are applicable to all cereals. Rye is somewhat of an exception, being an open-pollinated crop species. Plarre of the University of Berlin suggests that after a cross between two *Secale cereale* cultivars, the progenies should be allowed to open pollinate until the F_5 generation, when single plant selection is initiated.[26] This is followed by the first inbreeding which is actually half-sib mating in the F_6. Plarre conducted the first sprouting tests in the F_8 after the full-sib mating in F_7. Using their most promising line (strain 46) single and multiple crosses were made by putting varying numbers of plants together in a single bag to interpollinate. Plarre's team also screened rye species other than *Secale cereale* for preharvest sprouting resistance. They found that only *S. silvestre* could be used as a source of resistance. In crossbreeding, *S. silvestre* was used as male parent and from 3700 pollinated flowers, they obtained only two hybrid seeds. After backcrossing the two F_1 plants (used as male) with the female parent, they built up a pedigree system. This breeding technique is professed to be a typical model of introgressive hybridization. It is also alleged that, in the wild parent, the processes of germination and dormancy are controlled by dominant factors.[26] Nevertheless, Oettler does not see any reason to use *S. silvestre* as a source of sprouting resistance as the crosses between *S. cereale* and *S. silvestre* did not surpass the Swedish cv. Otello. 'Otello' itself has little importance in commercial production because of its low yield potential. Oettler argues that instead of using the wild species to improve the sprouting tolerance of rye, it is more advisable to utilize cultivated rye populations which already have some sprouting tolerance.[52]

IV. CONCLUSIONS

When an attempt is made to improve the preharvest sprouting resistance of a crop, it is only one of many breeding objectives. The improved crop should have high yield potential, appropriate quality, improved disease and pest resistance, lodging resistance, etc. It is no easy task to coordinate all these requirements, without losing some of the important attributes, due to the various selection pressures to which the population is exposed as well as other factors caused by intergenotypic competition, linkage relationships, etc.

Among the different strategies discussed, the backcrossing method is the safest but not necessarily the cheapest. Basically, the idea is to develop a genotype which is in every other respect acceptable except for sprouting resistance and then, by backcrossing, introduce this factor alone. One drawback in this strategy is that by the time the cultivar is ready for release, the pathogens, particularly rust, may have developed a new physiological strain to which the new cultivar is not resistant. With this in mind, it would appear safer to develop a basic cultivar with the appropriate yield potential, quality, and adequate preharvest sprouting tolerance and then add, by backcrossing, the disease resistance genes, as circumstances demand.

Considering the multiple factors governing preharvest sprouting resistance, a single cross between two parents will seldom be successful. To obtain an efficacious basic genotype on which further commercial cultivars can be based, a multiple crossing program is required whereby all traits are combined in such a manner so as to provide a high probability of successful recombinations. For each breeding location, the crossing program has to be specifically designed. No general prescription can be provided. The parents have to be carefully selected and the breeder must be familiar with the inheritance of the different traits to be combined. Nevertheless, these are the general rules of plant breeding and should be well known by every breeder.

The application of various selection methods may have a significant influence on the probability of recovering the required recombinations within an economically feasible time. The pedigree method in general has been acclaimed as the most accurate selection system.

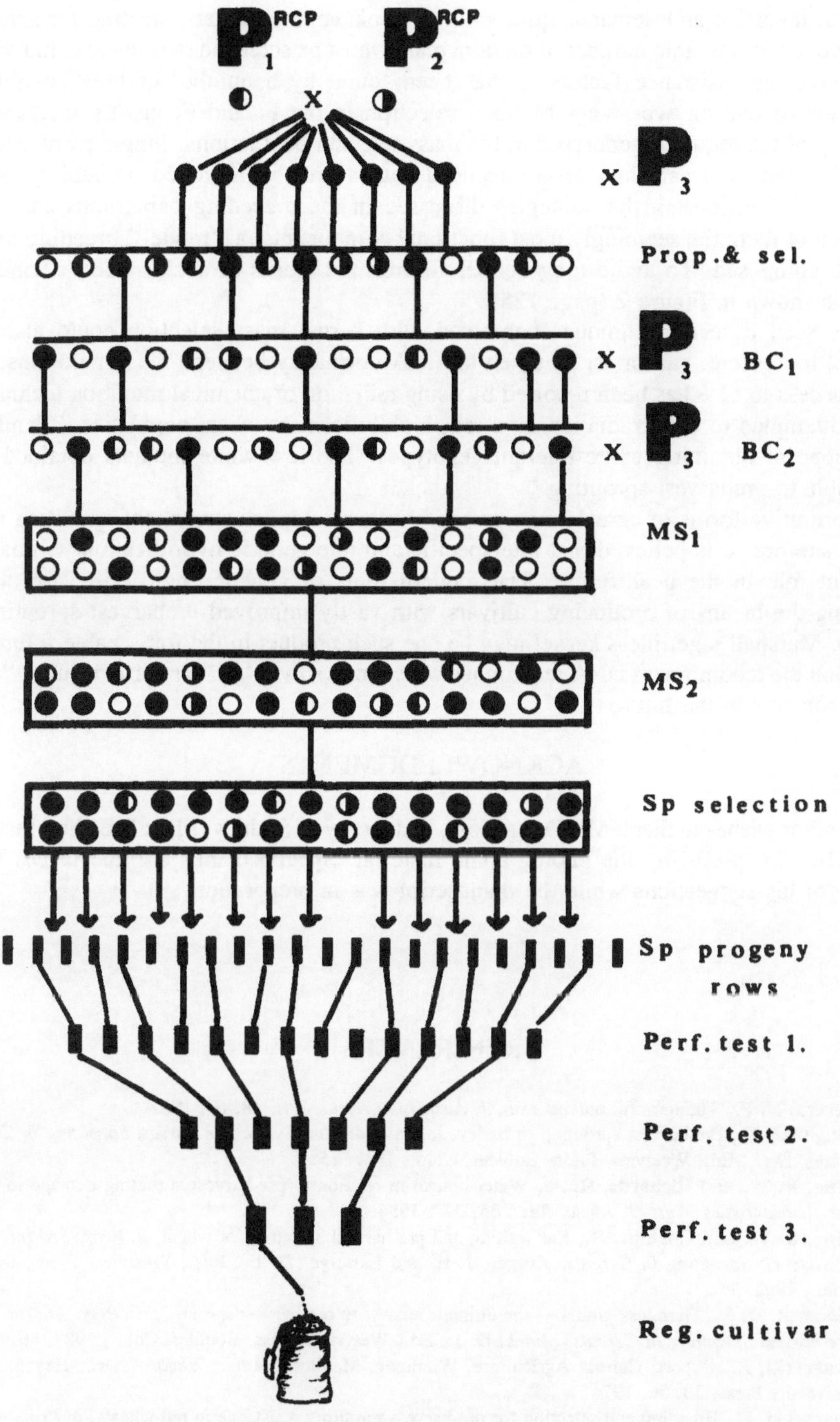

FIGURE 2. A "model" cereal improvement system, having preharvest sprouting tolerance as the major aim. P_1^{RCP} and P_2^{RCP} = parents representing two different sources of preharvest sprouting tolerance. P_3 = parent with high yielding, high quality, and disease resistant genes. Prop. & sel. = propagation and selection. BC = backcross. MS = mass selection. Sp = single plant. Perf. test = performance test. Reg. cultivar = registered new cultivar.

It is labor intensive and demands quite a lot of bookkeeping. When selecting for sprouting resistance, a considerable number of random plants must be screened to find those individuals which have the resistance factors. It has been found by a number of breeders that the application of one or two stages of mass selection in the F_2 and F_3 greatly increases the frequency of the required genotypes in the mass selected populations. Single plant selection in the F_4 population can then be commenced with a greatly improved probability of success.[1,40,45,53] Considering the strategies discussed in the preceding paragraphs and taking from each of them the seemingly most functional components, a "model" breeding method could be composed. To avoid long deliberation on the merits of such model, a schematic system is shown in Figure 2 (page 125).

Single seed descent techniques combined with kernel mass selection could also be a powerful tool. Here, care must be taken to break dormancy between the generations.

No success to date has been reported by using radiation or chemical mutation techniques. Derera attempted to use irradiation on a red, highly dormant wheat population ('Pembina') in the hope of obtaining white tolerant genotypes. The few white mutants obtained were susceptible to preharvest sprouting.[54]

The primitive forms of cereals may provide further untapped sources of sprouting resistance. Therefore, it is believed that interspecific and intergeneric hybridization will have an important role in the near future. The modern tools of genetic engineering are already promising the means of producing cultivars with vastly improved preharvest sprouting resistance. Marshall's germless kernel may be one such product in the foreseeable future, not to mention the recombinant DNA techniques which may solve "all" or perhaps just a "few" of the problems in the future.

ACKNOWLEDGMENTS

The author wishes to thank Mr. Doug Gow, University of Sydney, Plant Breeding Institute, Castle Hill, for preparing the photographic material. Special thanks are due to Dr. Norm Darvey for his suggestions while the manuscript was in preparation.

REFERENCES

1. **Derera, N. F.,** The harmful harvest rain, *J. Aust. Inst. Agric. Sci.*, 48, 67, 1982.
2. **Ringlund, K.,** Pre-harvest sprouting in barley, in *Proc. 4th Int. Symp. Pre-harvest Sprouting in Cereals*, Mares, D. J., Ed., Westview Press, Boulder, Colo., 1987, 15.
3. **King, R. W. and Richards, R. A.,** Water uptake in relation to pre-harvest sprouting damage in wheat: Ear characteristics, *Aust. J. Agric. Res.*, 35, 327, 1984.
4. **King, R. W. and Chadim, H.,** Ear wetting and pre-harvest sprouting of wheat, in *Proc. 3rd Int. Symp. Pre-harvest Sprouting in Cereals*, Kruger, J. E. and LaBerge, D. E., Eds., Westview Press, Boulder, Colo., 1983, 36.
5. **Marshall, D. R.,** Germless grains — the ultimate answer to pre-harvest sprouting, in *Proc. 4th Int. Symp. Pre-harvest Sprouting in Cereals*, Mares, D. J., Ed., Westview Press, Boulder, Colo., 1987, 86.
6. **Czarnecki, E.,** Report, Canada Agriculture, Winnipeg, Manitoba, *Annu. Wheat Newsl.*, Heyne, E. G., University Press, 20, 56, 1974.
7. **Czarnecki, E.,** Breeding and selection for pre-harvest sprouting resistance in red wheats, in *Proc. 4th Int. Symp. Pre-Harvest Sprouting in Cereals*, Mares, D. J., Ed., Westview Press, Boulder, Colo., 1987, 45.
8. **Olsson, G.,** Breeding for Sprouting Resistance in Wheat, mimeographed report to the Swedish Seed Association, Svalöf, 1974.
9. **McEwan, J. M.,** Breeding for resistance to pre-harvest sprouting in New Zealand wheats, *Cereal Res. Commun.*, 4, 97, 1976.
10. **Derera, N. F., McMaster, G. J., and Balaam, L. N.,** Pre-harvest sprouting resistance and associated components in 12 wheat cultivars, *Cereal Res. Commun.*, 4, 173, 1986.

11. **Linhares, A. G. and Dotto, S. R.,** Report of a study conducted in Brazil on the sprouting problem in wheat, *Cereal Res. Commun.*, 8, 251, 1980.
12. **McCaig, T. N. and DePauw, R. M.,** Falling numbers and α-amylase in sawfly-resistant wheats, in *Proc. 3rd Int. Symp. Pre-harvest Sprouting in Cereals*, Kruger, J. E. and LaBerge, D. E., Eds., Westview Press, Boulder, Colo., 1983, 79.
13. **DePauw, R. M., McCaig, T. N., and Townley-Smith, T. F.,** Registration of a sprouting resistant white-seeded spring wheat germplasm line, *Crop Sci.*, 25, 577, 1985.
14. **Bhatt, G. M., Paulsen, G. M., Kulp, K., and Heyne, E. G.,** Pre-harvest sprouting in hard winter wheats: Assessment of methods to detect genotypic and nitrogen effects and interactions, *Cereal Chem.*, 58, 300, 1981.
15. **Moss, H. J. and Kirby, A.,** A role for fibrous material in flour paste viscosity of wheat, *Cereal Res. Commun.*, 4, 221, 1976.
16. **Mares, D. J., Ellison, F. W., and Derera, N. F.,** Gibberellic acid insensitivity genes and pre-harvest sprouting damage resistance, in *Proc. 3rd Int. Symp. Pre-harvest Sprouting in Cereals*, Kruger, J. E. and LaBerge, D. E., Eds., Westview Press, Boulder, Colo., 1983, 22.
17. **Mares, D. J.,** Pre-harvest sprouting tolerance in white-grained wheat, in *Proc. 4th Int. Symp. Pre-harvest Sprouting in Cereals*, Mares, D. J., Ed., Westview Press, Boulder, Colo., 1987, 64.
18. **Mares, D. J.,** Investigation of the pre-harvest sprouting damage mechanisms in some Australian white wheats, in *Proc. 3rd Int. Symp. Pre-harvest Sprouting in Cereals*, Kruger, J. E. and LaBerge, D. E., Eds., Westview Press, Boulder, Colo., 1983, 59.
19. **King, R. W.,** Water uptake in relation to pre-harvest sprouting damage in wheat: Grain characteristics, *Aust. J. Agric. Res.*, 35, 337, 1984.
20. **Gale, M. D. and Marshall, G. A.,** Insensitivity to gibberellin in dwarf wheats, *Ann. Bot.*, 37, 729, 1973.
21. **Gale, M. D.,** Towards super cereals, *New Sci.*, January 31, 1974.
22. **Harvey, B. L., Rossnagel, B. G., and Muderewich, R. P.,** Sprouting resistance in barley, in *Proc. 3rd Int. Symp. Pre-harvest Sprouting in Cereals*, Kruger, J. E. and LaBerge, D. E., Eds., Westview Press, Boulder, Colo., 1983, 239.
23. **Reitan, L.,** Early generation mass selection for seed dormancy in barley, in *Proc. 3rd Int. Symp. Pre-harvest Sprouting in Cereals*, Kruger, J. E. and LaBerge, D. E., Eds., Westview Press, Boulder, Colo., 1983, 244.
24. **Jonassen, I.,** Segregation of high and low chymotrypsin inhibitor CI-2 content in crosses between Hiproly and normal lysine barley genotypes, *Carlsberg Res. Commun.*, 47, 305, 1982.
25. **Persson, E.,** Otello - a result of amylase selection for sprouting resistance, *Cereal Res. Commun.*, 4, 101, 1976.
26. **Plarre, W.,** Breeding methodology for rye with resistance to sprouting, *Cereal Res. Commun.*, 8, 265, 1980.
27. **Piech, J., Ruszkovski, M., and Jaworska, K.,** Inheritance of seed dormancy stage duration in winter wheat (*Triticum aestivum* L.), *Genet. Pol.*, 11, 227, 1970.
28. **DePauw, R. M. and McCaig, T. N.,** Recombining dormancy from RL 4137 with white seed color, in *Proc. 3rd Int. Symp. Preharvest Sprouting in Cereals*, Kruger, J. E. and LaBerge, D. E., Eds., Westview Press, Boulder, Colo., 1983, 251.
29. **Derera, N. F.,** Should red wheat be introduced into the northern wheatbelt, *J. Aust. Inst. Agric. Sci.*, 39, 48, 1973.
30. **Bhatt, G. M., Derera, N. F., and McMaster, G. J.,** Breeding white-grained spring wheat for low alpha-amylase synthesis and insensitivity to gibberellic acid in grain, *Cereal Res. Commun.*, 4, 245, 1976.
31. **McMaster, G. J. and Derera, N. F.,** Methodology and sample preparation when screening for sprouting damage in cereals, *Cereal Res. Commun.*, 4, 251, 1976.
32. **Derera, N. F. and Bhatt, G. M.,** Germination inhibition of the bracts in relation to pre-harvest sprouting tolerance in wheat, *Cereal Res. Commun.*, 8, 199, 1980.
33. **Bhatt, G. M. and Derera, N. F.,** Potential use of Kenya 321-type dormancy in wheat breeding programme aimed at evolving varieties tolerant to pre-harvest sprouting, *Cereal Res. Commun.*, 8, 291, 1980.
34. **Mares, D. J.,** Comparisons of tolerance to pre-harvest sprouting within *Triticum* species, triticale and barley, in *Proc. 4th Int. Symp. Pre-harvest Sprouting in Cereals*, Mares, D. J., Ed., Westview Press, Boulder, Colo., 1987, 148.
35. **Hejgaard, J. and Gibbons, G. C.,** Screening for α-amylase in cereals. Improved gel-diffusion assay using dye-labeled starch substrate, *Carlsberg Res. Commun.*, 44, 21, 1979.
36. **DePauw, R. M. and McCaig, T. N.,** Recovery of sprouting resistance from red-kerneled wheats in white-kerneled segregates, in *Proc. 4th Int. Symp. Pre-harvest Sprouting in Cereals*, Mares, D. J., Ed., Westview Press, Boulder, Colo., 1987, 54.
37. **Derera, N. F., Bhatt, G. M., and McMaster, G. J.,** On the problem of pre-harvest sprouting of wheat, *Euphytica*, 26, 299, 1977.

38. **Derera, N. F. and Bhatt, G. M.,** Effectiveness of mechanical mass selection in wheat (*Triticum aestivum* L.), *Aust. J. Agric. Res.*, 23, 761, 1972.
39. **Derera, N. F., Bhatt, G. M., and Ellison, F. W.,** Population improvement through mechanical mass selection, *Cereal Res. Commun.*, 2, 95, 1974.
40. **Svensson, G.,** Screening methods for sprouting resistance in wheat, *Cereal Res. Commun.*, 4, 263, 1976.
41. **Fukunaga, K., Hoshino, T., Matsukura, U., Taira, H., and Oda, S.,** Differences in pre-harvest sprouting and alpha-amylase activity among wheat cultivars, in *Proc. 4th Int. Symp. Pre-harvest Sprouting in Cereals*, Mares, D. J., Ed., Westview Press, Boulder, Colo., 1987, 116.
42. **Gale, M. D., Flintham, J. E., and Arthur, E. D.,** Alpha-amylase production in the late stages of grain development — an early sprouting damage risk period, in *Proc. 3rd Int. Symp. Pre-harvest Sprouting in Cereals*, Kruger, J. E. and LaBerge, D. E., Eds., Westview Press, Boulder, Colo., 1983, 29.
43. **Fischnich, O. and Thielebein, M.,** Hinweise für den Mähdreschereinsatz 1954, *Mitt. DLG.*, 69, 545, 1954.
44. **Weilenmann, F.,** Beitrag zur Prüfmethodik in der Züchtung auf Auswuchsfestigkeit bei Weizen (*Triticum aestivum* L.), Ph.D. Diss. No. 5712, Eidgenössischen Technischen Hochschule, Zürich, 1976.
45. **Strand, E.,** A seed dormancy index for selection of cultivars of cereals resistant to pre-harvest sprouting, *Cereal Res. Commun.*, 8, 219, 1980.
46. **Larsson, S.,** Selection for seed dormancy by using germination tests, in *Proc. 4th Int. Symp. Pre-harvest Sprouting in Cereals*, Mares, D. J., Ed., Westview Press, Boulder, Colo., 1987, 400.
47. **Weilenmann, F.,** A selection method to test the sprouting resistance in wheat, *Cereal Res. Commun.*, 4, 267, 1976.
48. **Weilenmann, F.,** Plant breeding aspects of sprouting resistance with the selection technique using sprouting index, *Cereal Res. Commun.*, 8, 209, 1980.
49. **Weilenmann, F.,** Ten years experience using sprouting index in wheat breeding, in *Proc. 4th Int. Symp. Pre-harvest Sprouting in Cereals*, Mares, D. J., Ed., Westview Press, Boulder, Colo., 1987, 93.
50. **Strand, E.,** Effects of temperature and rainfall on seed dormancy of small grain cultivars, in *Proc. 3rd Int. Symp. Pre-harvest Sprouting in Cereals*, Kruger, J. E. and LaBerge, D. E., Eds., Westview Press, Boulder, Colo., 1983, 260.
51. **Gordon, I. L.,** Factor analyses of characters useful in screening wheat for sprouting damage, in *Proc. 3rd Int. Symp. Pre-harvest Sprouting in Cereals*, Kruger, J. E. and LaBerge, D. E., Eds., Westview Press, Boulder, Colo., 1983, 221.
52. **Oettler, G.,** Pre-harvest sprouting in rye and triticale, in *Proc. 4th Int. Symp. Pre-harvest Sprouting in Cereals*, Mares, D. J., Ed., Westview Press, Boulder, Colo., 1987, 24.
53. **Reitan, L.,** Breeding and selection for pre-harvest sprouting tolerance in barley, in *Proc. 4th Int. Symp. Pre-harvest Sprouting in Cereals*, Mares, D. J., Ed., Westview Press, Boulder, Colo., 1987, 75.
54. **Derera, N. F.,** unpublished.

Chapter 7

PREHARVEST SPROUTING DAMAGE AND SPROUTING TOLERANCE: ASSAY METHODS AND INSTRUMENTATION

Daryl J. Mares

TABLE OF CONTENTS

I. INTRODUCTION AND SCOPE

Preharvest sprouting represents a major constraint to the production of quality grain in many parts of the world and accordingly it has been the subject of a considerable amount of research. In a few areas the severity of the problem has been considerably reduced by improved agronomic management or breeding; however, in most sprouting-prone regions a practical solution has not yet been achieved. Implementation of crop management strategies, such as early harvest followed by artificial drying,[1e] which reduce the chance of rain falling on the mature crop, can reduce the incidence and severity of preharvest sprouting damage. Where damage has occurred, sophisticated, objective testing systems have been developed in most countries to ensure that damaged grain is detected at receival and correctly graded and segregated.[2-4] In addition, there is some scope for salvaging good quality grain from partially or lightly damaged samples by physical means,[5,6] for blending small quantities of damaged grain with sound grain,[7] for directing lightly damaged grain into products or manufacturing areas which are less sensitive to the biochemical changes which occur during sprouting,[8] and for modifying process technology to counteract some of the effects of sprout damage.[7] All these strategies for identification, separation, and optimum utilization of sprout-damaged grain require accurate methods for the quantitative assessment of preharvest sprouting damage both in the field and in the laboratory. The data must be acceptable and meaningful to growers, grain handling authorities, grain marketing bodies, and grain processors.

Segregation and salvage operations involve an additional cost factor which must be added to the cost of production and product manufacture. Another approach to the problem of preharvest sprouting in cereals is to develop cultivars which are more tolerant to adverse weather conditions at maturity. Reliable and efficient methods for screening cultivars and selecting within breeding populations for sprouting tolerance are required for this work.

This chapter will discuss the methods, techniques, and equipment which can be used to:

1. Assess the tolerance of breeding lines and established cultivars to sprouting,
2. Determine the likely mechanisms responsible for tolerance, and
3. Assess the level of sprouting damage in the laboratory and in the field.

Discussion will be restricted to precocious grain germination in the ear in response to rain and the biochemical changes which ensue, and will not consider secondary effects which may result, for example, from infestation of weathered grain by microorganisms. The impact of preharvest sprouting damage is most acute in wheat because a high proportion of this important crop is destined for human consumption and in most instances the quality and the appearance of wheat flour products are adversely affected by even relatively minor levels of damage.[8] As a consequence, more research has been directed to wheat than the other temperate cereals and the majority of methods have been specifically developed for wheat grain and flour. Most of the methods can, however, be adapted to other cereals, particularly barley, rye, and triticale, and, presumably, also to oats and rice.

II. SELECTION FOR TOLERANCE TO PREHARVEST SPROUTING

The emphasis which is placed on the development of sprouting-tolerant cultivars will depend on the likely incidence and severity of sprouting in the geographical region concerned, the economic effects on grain producers, and the magnitude of the problems which arise in the marketing and processing sections of the industry. Sprouting tolerance must be considered in relation to other agronomic and disease problems and the likelihood of making a significant advance in the face of the prevailing harvest weather conditions and the genetic variability available to the breeders. Breeding and selection methodologies are discussed by Derera

(Chapter 6) and in numerous papers presented at the 1st, 2nd, 3rd, and 4th International Symposia on Pre-Harvest Sprouting in Cereals.[1a-d]

A. Definitions, Selection Criteria, and Plant Culture

Preharvest sprouting tolerance can be defined as a reduced rate of germination of grain in the ear and a reduced rate of production of germinative enzymes under conditions normally conducive to sprouting. Preharvest sprouting damage refers to the changes in the grain constituents, such as hydrolytic enzymes, which affect the processing properties of the grain or the viability and germinability of the grain under long-term storage. Methods for screening breeding lines and cultivars or selecting tolerant individuals from breeding populations need to be considered in relation to a number of criteria. These include cost, sample size, labor, reliability and reproducibility, ability to discriminate between good and poor cultivars, and capacity. For the accurate characterization of tolerant lines and cultivars, more emphasis will have to be placed on objective physical and chemical tests. In addition to a suitable testing program, there needs to be a reliable recording of daily weather conditions, particularly during grain filling and maturation, since environmental factors such as temperature and rainfall influence tolerance and the interpretation of data.

Where possible, plant material should be grown in field trials, avoiding abnormal sowing times, soil types, and agronomic practices, and duplicating as nearly as possible local culture methods. In the author's experience, well-spaced small plots (twin rows x 2 to 5 m long) in a random block design give consistent and reliable results which correlate with the performance of lines under commercial cultivation. Larger plots (10 rows × 10 to 25 m) show more variation in maturity and head size, but nevertheless show substantially similar rankings to smaller plots. The use of strong healthy plants helps reduce errors due to stress and disease. Greenhouse-grown plants may be useful in particular circumstances but, in general, the degree of expression of tolerance does not relate very well to that in the field-grown material.[9]

B. Assessment of Crop Maturity

Grain development, maturation, and after-ripening are not discrete events in the life of a cereal grain, but rather they simply represent recognizable stages in a continuous dynamic process. Gordon has reported that various grain characteristics change at different rates during this process depending on genotype and environment.[10,11] For example, under cool ripening environments (15 to 18°C), embryo maturation was slowed to a greater extent than grain dehydration. Tolerance to preharvest sprouting and the extent of damage caused by a particular episode of unfavorable weather will depend to a large extent on the stage of crop maturation (see changes in grain germinability during grain development described by Gosling et al.[12] and Mares[13]). There is a requirement, therefore, for the accurate assessment of the sprouting tolerance of different breeding lines and cultivars, to monitor grain maturity, to establish a reference point for comparing cultivars and trials, and to reduce the risk of selecting for late-maturing genotypes. The reference point may be morphological maturity (i.e., attainment of maximum grain dry weight), harvest maturity, or some other conveniently identified stage. A particular maturity or ripeness indicator may not necessarily be universally applicable because of genotype-environment interactions. In some instances, even the definition of the maturity index may differ from country to country. Thus, harvest ripeness in Australia refers to the maximum permitted moisture content for wheat grain delivery of 12%, whereas in other countries, with less favorable drying conditions, the permitted moisture content may be considerably higher (Canada 13.5%, Sweden 15%). Theoretically, harvest ripeness simply implies readiness for harvest, with the grain being firm enough to escape damage under the particular harvest method being used. Despite this variation, there is a strong case for using harvest ripeness, or an index of harvest ripeness, because of its special relevance to sprouting (in practice), to growers, and to the harvest and delivery of grain.

The methods which have been applied to the assessment of crop maturity can be divided into two groups:

1. Subjective methods, e.g., loss of green pigments from the glumes or flag leaves, grain color, grain firmness, appearance of the pigment strand in the grain
2. Objective methods, e.g., grain dry weight, grain moisture content, movement of dye-labeled water into the stem and vegetative structures of the ear

The loss of green pigments from vegetative plant parts is an attractive method, particularly for large quantities of material, since it is quick and nondestructive. Copeland and Crookston showed that the loss of grain color from the glumes and peduncle of barley ears coincided with maximum grain dry weight, whereas the pigment strand appeared 1 to 4.5 days later and loss of flag leaf color coincided with 95% maximum kernel dry weight.[14] Hanft and Wych reported similar findings for hard red spring wheat.[15] Kernel moisture generally appears to be too variable in relation to the time of attainment of maximum grain dry weight. Several investigations involving barley and wheat have indicated that kernel moisture at maximum grain dry weight is influenced by environmental conditions and may range from 32 to 44% in wheat and 18 to 39% in barley.[14-16] Clarke observed little difference among cultivars in the rate of water loss after attainment of maximum grain dry weight.[16] The subjective assessment of kernel maturity by a "pinchtest" is a useful additional method which may also be applied to situations where the results of other methods are equivocal.[9] Periods of high temperature (>30°C) during grain filling may cause the loss of color from flag leaves and stems to be markedly inhibited relative to ear and grain dehydration. Similarly, severe infection with diseases such as stem rust can cause premature grain dessication.

Eosin ripeness (i.e., cessation of movement of water labeled with the red dye eosin into the vegetative structures of the ear) has been used for several years in the author's laboratory and several other institutions and has proven to be a useful method for comparing cultivars in a particular trial.[17] The method is time consuming and destructive of material and, therefore, not readily applicable to large numbers of samples or breeding populations. With this method, ears are harvested at random from field plots (5 to 7 ears per plot), leaving at least one stem node attached, and transported to the laboratory. The material is recut under water between the ear and the first node, placed in a 1% aqueous solution of eosin, and left for 3 hr. Ears are scored according to whether red dye has moved into the stem, into the rachis, and into the glumes. Tests are repeated at intervals until eosin ripeness occurs, i.e., when red dye no longer moves into the stems of the harvested ears. With experience, the number of tests required can be substantially reduced, and once the pattern in a particular trial has been established, it can be replaced by some other maturity index, such as loss of green color, without significant loss in accuracy.

In large plots, maturity assessment may be complicated by variation in plot density, soil moisture status, edge effects, and crop architecture. Tiller mortality following frost damage or severe stress may result in the activation of tertiary tillers giving rise to two distinct maturity stages within the plots. In the author's experience, little useful information on comparative sprouting tolerance can be derived from such material.

C. Sampling and Storage of Plant Material

The response of intact ears of cereals in the field to preharvest rain includes effects conditioned by the presence of the vegetative parts of the ear and by the integrity of the layers of maternal reproductive tissue which form the seed coat. Consequently, care must be taken during harvesting to preserve these structures in their native state as far as possible. Hand harvest of individual ears at random is recommended. Ears should be taken from

Table 1
EFFECT OF AN ARTIFICIAL WETTING TREATMENT (WD) ON THE FALLING NUMBER (FN) OF WHEAT SAMPLES HARVESTED AT EOSIN RIPENESS (ER) OR AT 5 DAYS AFTER EOSIN RIPENESS (ER+5)

		Falling number (sec)					
		5 days after ER		15 days after ER		25 days after ER	
Variety	Harvest	IFN	WD-FN	IFN	WD-FN	IFN	WD-FN
Kite	ER	452	432	460	358	475	275
	ER+5	458	425	449	365	470	250
Songlen	ER	410	349	430	215	425	95
	ER+5	419	328	422	204	432	102

Note: Samples stored under cover at ambient temperature after harvest were tested at 5, 15, and 25 days after ER. (IFN = falling number before wetting treatment.) SE = ±25 sec.

primary tillers, where possible, although Salmon and Helm have observed that primary, secondary, and tertiary tillers of spring triticale produce similar rankings of lines and cultivars.[18] The time of harvest is dependent on individual requirements and experiences. In the author's program, material is harvested at eosin ripeness (15 to 20% moisture) and stored at ambient temperature, under cover and protected from rodents, until required. Under conditions normally experienced at harvest in northern N.S.W., the moisture content of this material declines to approximately 12% (harvest maturity) in 5 days. By contrast, Czarnecki prefers to harvest wheat material in Manitoba, Canada at around 25% moisture and dry it to 13.5% prior to testing.[9] A comparison of material harvested at eosin ripeness and stored under cover for 5 days with material left in the field for 5 days after eosin ripeness did not reveal any significant differences in the response to artificial wetting treatments (Table 1). In parts of Canada, wheat is swathed when it reaches 35% moisture, to promote dehydration, and left in windrows in the field. Clarke has shown that swathed material is more susceptible to sprouting than identical material left standing in the field.[19]

Harvest-mature ears may be stored for some months in a deep freeze at −20°C without affecting the response to wetting treatments (Tables 2 and 3). Similarly, grain from harvest-mature ears may be stored for at least 6 months at −20°C.[20,21] Provided that the moisture content is approximately 15% or less, there seems to be little risk of tissue damage by freezing and subsequent thawing.[21] Grain required for the study of postmaturity changes in germinability should be stored so as to avoid drastic fluctuations in humidity, unless the effect of this factor is actually the subject of study.

The threshing of ear samples to obtain grain for germinability or dormancy tests, and indeed for later biochemical investigations, should be performed with care, and methods such as hand-rubbing or hand-dissection are recommended. Mechanical methods may damage the pericarp and testa layers covering the grain or even the embryo itself. In a study involving a mechanical single-plant thresher (Walter and Wintersteiger Gsmbh., Austria) short threshing times (1 to 2 sec) resulted in a marked increase in germinability, whereas progressively longer periods resulted in decreased germinability and an increasing proportion of nongerminable grains (Figure 1). These results can be explained in terms of loss of dormancy in the shorter treatment periods due to damage to the pericarp and damage to the germ tissues with longer treatment periods.

Table 2
RESPONSE TO A STANDARD WETTING TREATMENT ON WHEAT EARS STORED AT AMBIENT TEMPERATURE (20 TO 36°C) OR −20°C FROM HARVEST MATURITY

	Falling number (sec)					
	Days storage at ambient temp.				Days storage at −20°C	
Variety	0	10	20	50	20	50
Cook	420	402	206	95	393	431
Songlen	428	373	171	65	429	448
Kite	453	431	438	350	461	459

Table 3
RESPONSE TO A STANDARD WETTING TREATMENT ON WHEAT EARS STORED AT AMBIENT TEMPERATURE (25 TO 36°C) AFTER HARVEST AND TRANSFERRED TO −20°C STORAGE AT 5, 15, AND 25 DAYS AFTER EOSIN RIPENESS FOR 1 MONTH

	Falling number (sec)					
	5 days		15 days		25 days	
Variety	C	−20°C	C	−20°C	C	−20°C
Timgalen	265	250	150	124	65	73
Songlen	103	130	98	101	63	70
Shortim	333	309	299	274	104	124

Note: Control treatments (C) refer to wetting treatments applied to samples at the time of transfer to −20°C.

D. Selection and Screening Methods

The propensity to sprout in the ear depends on characteristics associated with the vegetative structures of the ear, e.g., transfer of water and gases between the environment and the surface of the grains, presence of germination inhibitors, and grain characteristics, e.g., uptake of water, transfer of gases, and grain germinability. Sprouting tolerance has often been regarded as synonymous with grain dormancy. However, while it is clear that this is a very important component of tolerance,[22] King and Richards[23] and Mares[24] have identified physical mechanisms associated with the vegetative parts of the wheat ear, which appear to give some extra tolerance. Other workers have proposed that germination inhibitors in the bracts or chaff of wheat ears may be partly responsible for sprouting tolerance.[25,26] Where grain germinability is considered to be the most important factor, then simple germination tests should serve as adequate selection and screening methods.[27-29] Alternatively, sprouting in intact ears, in response to natural or artificial wetting, incorporates all the possible ear factors and avoids both the labor involved in threshing and the possibility of altering grain germinability.

1. Grain Germination Tests

Specific techniques used to estimate grain dormancy and grain germinability will be

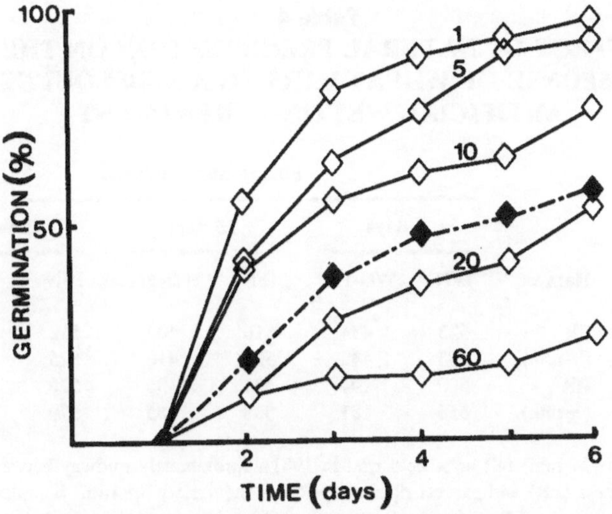

FIGURE 1. Germination at 20°C of wheat grain threshed for varying times in a single plant thresher (◇) (Walter and Wintersteiger, Austria) compared with grain dissected from ears by hand (◆).

discussed in a later section of this chapter. Noll et al.,[27] Larsson,[29] and Reitan[30] have described the use of grain germination tests to select for sprouting tolerance in wheat and barley breeding programs, respectively. In these instances germination tests were performed at harvest maturity. Noll and Czarnecki described two techniques for extending the testing period and thereby greatly increasing the quantity of material which could be screened.[20] The first method, already referred to, involved preservation of harvest-mature grain germinability by deep-freeze storage. This technique has also been investigated by Mares, who further showed that even after 3 to 6 months at −20°C, grain which was returned to +12°C storage afterripened at an identical rate to grain which had been placed in +12°C storage at harvest ripeness. The second method involved germination at 30°C, rather than the more usual 18 to 20°C, where dormancy is more strongly expressed.[12,13,31] Grain, which had been stored at an ambient temperature for some weeks after harvest ripeness and had consequently lost a considerable portion of its dormancy, germinated readily at 18°C but at 30°C germination was still partially inhibited. The magnitude of this residual effect was found to be related to the degree of dormancy shown at harvest.[20] This technique is only useful for a limited period after maturity since even the high temperature-induced reduction in germinability gradually disappears with time. In another study, the high temperature effect could not be detected after 6 months storage at 12°C.[13] Germination at the higher temperature has also been used to deliberately enhance dormancy and assist selection in populations where dormancy is inherently low and where conditions have been unfavorable for development of good dormancy.[18]

2. Artificial Wetting Treatments

Efficient selection and breeding for preharvest sprouting tolerance requires (1) accurate differentiation between lines and (2) the capability to apply a screening test every season, irrespective of the prevailing weather pattern, to large numbers of individual lines. The obvious strategy which meets these conditions is to treat ear samples at identical stages of maturity with a standard wetting treatment and to quantify the effects of this treatment by visual or objective means. This treatment can be repeated at intervals after ripeness, thus giving an estimate of tolerance at harvest ripeness and the duration tolerance, and to enable missing values to be calculated by extrapolation. Alternatively, the wetting-treatment period

Table 4
EFFECT OF NATURAL PRECIPITATION ON THE RESPONSE OF WHEAT EARS TO A SUBSEQUENT, ARTIFICIAL WETTING TREATMENT

Replicate	Harvest	Falling number (sec.) 5 days IFN	5 days WD-FN	15 days IFN	15 days WD-FN	25 days IFN	25 days WD-FN
1	ER	523	434	516	405	512	443
2	(−rain)	505	442	535	410	525	403
3	HR	500	194	531	135	526	83
4	(+rain)	556	181	538	155	529	101

Note: Rain (24 mm) fell on a field trial in 1981 approximately midway between eosin ripeness (ER) and harvest ripeness (HR) for the variety Shortim. Samples of ears were harvested from four replicate plots at ER or HR, stored under cover at ambient temperature, and subjected to a *standard wetting treatment* at 5, 15, or 25 days after ER. (IFN = falling number prior to treatment, WD-FN = falling number after wetting treatment).

can be extended and serial time samples used to estimate the duration of the wetting treatment required to overcome the tolerance.

The occurrence of rain during the harvest of varieties in a particular trial may make the interpretation of results extremely difficult. Rain at or near harvest ripeness may not be sufficient to cause visible signs of damage but may make the sample much more susceptible to the subsequent artificial wetting treatment (Table 4). To some extent this complication can be overcome by including cultivars of known tolerance, but with a range of maturities, in the trials. In the author's program, three white spring wheat cultivars (Eagle, Sunelg, and Kite), which have similar levels of tolerance all derived from the same source but with an approximately 7- to 10-day spread in maturity, are included in all trials as standards.

A number of wetting treatments have been described in the literature. These include:

1. A rain simulator, based on a Svalov design, in which overhead misty spray is applied to spaced spikes held erect on perforated trays rotating at uniform speed in a chamber maintained at high relative humidity.[32]
2. A modification of the above system which used direct spray (50 mm in 2 hr) and accurate temperature control (20°C ± 1°C) in addition to high relative humidity (Figure 2)[22]
3. A misting chamber with facilities to vary intensity of misting[23,33]
4. A Seeger micromalter unit in which a relatively small number of ears can be misted for 10 sec every 30 min for a total of 48 hr at different temperatures[34]
5. Submersion of ears in water for a specified period

The period of treatment varies considerably between laboratories and depends on the cultivars, the stage of maturity, the level of tolerance at harvest ripeness, and, of course, the methods which will be used to assess tolerance or damage. Trial treatments to establish the optimum time for good discrimination between varieties are suggested. The control of temperature is important in view of its pronounced effect on grain germinability.

There would appear to be some merit in attempting to simulate the intensity of rainfall typical of the harvest period in the region concerned. King and Richards examined the influence of different rates of misting from 2 mm/hr to 8.4 mm/hr.[23] Wetting rate increased

FIGURE 2. Artificial rain simulation system in use at the Plant Breeding Institute, Narrabri. (M = mister, A = air conditioner.)

with rate of misting up to approximately 8 mm/hr and was comparable with natural rainfall. The higher rates of misting gave a more even treatment over a large area and did not appear to adversely affect cultivar differences in ear sprouting. Artificial wetting methods have subsequently been reviewed by King.[33] He pointed out that water droplet size, water droplet size distribution, and droplet velocity in these systems generally were not comparable with natural rain. However, he also conceded that, due to technical problems, it would be extremely difficult to simulate natural rainfall exactly at all or any particular geographic location. In the author's laboratory, misting rates of 25 mm/hr are used to simulate the very heavy rain characteristic of summer thunderstorms. Under this misting intensity, ears reached 70 to 100% of saturation, depending on cultivar, in 30 min and close to 100% saturation in 2 hr in all cultivars tested. In a comparative study conducted at Narrabri, four wetting treatments were applied to 11 wheat cultivars showing a range of tolerance to preharvest sprouting:

1. 50 mm of rain in 2 hr
2. 50 mm of rain in 2 hr followed by 6 mm every 24 hr.
3. 12.5 mm of rain in 30 min
4. Ears submerged in water for 2 hr

In all treatments, the ears were subsequently maintained at high humidity and 20°C. Rank according to sprouting tolerance was not significantly affected by the different treatments;

however, rate of appearance of seminal roots from spikelets was retarded in treatment 3 and slightly accelerated in treatments 2 and 4. Variation within cultivars was greatest for treatment 3.

The number of ears tested per line will depend to a large extent on the amount of material, the accuracy required, and the method of assessment to be used. For breeding populations, a small number of ears per individual plant or a bulked sample may be tested and the most susceptible lines discarded with no attempt being made to rank individual lines or ears. At the other extreme three replicates of ten ears give sufficient grain, when bulked after treatment, for an α-amylase assay and one or two falling number determinations. The system in operation at Narrabri consists of two identical chambers (see one section illustrated in Figure 2) which can handle approximately 750 lines (10 ears per line) per week, where visual sprouting is used to rank lines, or approximately 400 lines (3 replications of 10 ears per line) per week where falling number determinations are required.

3. Natural Rain

Generally speaking, natural precipitation is unpredictable in rate, duration, and occurrence. Nevertheless, some useful information can be obtained if these deficiencies are taken into account and efforts are made to monitor crop maturation and to collect daily meteorological data. Comparative information on the performance of cultivars in the field and in artificial wetting treatments may give some insight into the effects of crop microenvironment and crop canopy architecture on sprouting. In artificial rain simulation systems, differences due to plot density, position of ears in relation to leaves, and ear angle are normally eliminated. Preliminary data obtained in the author's laboratory indicate that the cultivar Shortim, a *Rht2* dwarf wheat in which the ears barely protrude above the leaf canopy, performs rather better in the artificial wetting system than in the field. By contrast, tall wheats, with long stems holding the ears well above the flag leaves, sometimes perform rather better in the field. These results could be explained in terms of a higher humidity being maintained at the level of the leaf canopy compared with a good drying environment for the ears of tall wheats. Experience over several years at Narrabri in northern N.S.W. has shown that lines selected for tolerance using the artificial wetting treatment have performed very well under commercial conditions. This generates a considerable degree of confidence in the testing system.

4. Assessment of Sprouting Tolerance

Sprouting or sprouting damage caused by artificial or natural weather treatments can be assessed by a number of subjective (visual ear-sprouting rate, visual ear-sprouting score, examination of threshed grain for evidence of germination) or objective (α-amylase activity, falling number, etc.) methods. Details of a number of objective methods will be discussed in Section IV. of this chapter. Of these, falling number and direct assay for α-amylase activity, are widely used in breeding and selection programs.

McMaster and Derera described the use of a sprouting score to rank cultivars following a fixed period of artificial wetting.[32] Ears were graded from 1 (no visible sprouting) through 5 (seminal roots uniform over spikes but not coleoptiles) to 11 (fully sprouted — coleoptiles uniform over spike and entering first leaf stage), and a mean was calculated for each line. The period of treatment may be altered to give optimum differentiation between good and poor cultivars. A number of factors such as fungal and bacterial infestation of ears, sterility of ears, variability in maturity stage, and prior weather conditions may affect the reproducibility of this method. In the author's limited experience with sprouting scores, problems were encountered at the more tolerant end of the spectrum. During daily observation, ears of susceptible genotypes showed an orderly and sequential progression through stages 1 to 11 (Figure 3a,b) as outlined by McMaster and Derera.[32] In contrast, more tolerant lines were often characterized by one or two sprouted grains, which subsequently developed shoots,

FIGURE 3. Sprouted ears of wheat. (a and b) Ears of susceptible cultivars showing even sprouting along the length of the spike; (c) ear of a sprouting-tolerant cultivar with only one badly sprouted grain.

on an otherwise unsprouted ear (Figure 3c). Such cultivars were difficult to score. Nevertheless this system has been widely accepted throughout the world and deserves serious consideration.

In the author's laboratory, daily observations of visible sprouting are made and ears discarded when seminal roots appear from one or more spikelets.[22] Observations are continued until all ears have sprouted and a relationship between percent sprouted ears and treatment time is established. Ears which do not sprout in 10 days are checked for seed viability. This method gives consistent discrimination between lines over a wide range of tolerance, requires somewhat less time than the sprouting score, and gives a good indication of the variation within a sample. The method has been successfully used to compare a large number of bread wheats, wheat relatives (hexaploids, tetraploids, and diploids), barleys, and triticales.[35] A study in which both methods were used to compare 25 white- and red-grained wheats grown at Narrabri indicated that similar rankings could be obtained with both systems.

Visual methods are particularly suitable for large numbers of lines or populations where sample size is limiting or where the aim is simply to select the most tolerant individuals. In the latter situation, the wetting treatment is continued until 70% of the ears have sprouted,[9] or until the response of standard cultivars indicates that a reasonable level of tolerance has been reached. Grain from nonsprouted ears can be retained for increase, further crossing, or further selection. Visual methods may also be used to confirm or support results obtained with objective methods such as falling number. Gordon has reported that low germination in a rain simulator has a high heritability and is therefore a good selection procedure.[36]

At some stage in cultivar development it is important to check α-amylase activity, falling number, and possibly amylograph peak viscosity, and the response of these parameters to weathering, since visual sprouting is not necessarily related to the inherent viscosity of the grain components or the activity of α-amylase. Some wheat lines (Figure 4) and most rye and triticale cultivars are characterized by low falling numbers and high α-amylase activities at maturity even in the absence of unfavorable weather conditions.[22,37-39] In the hard, white wheat-breeding program at Narrabri, 3 × 10 ears per line, hand-harvested at random from field plots (Figure 5), are loaded onto perforated aluminum trays (Figure 6, see also Figure 2) with sufficient stem to hold ears approximately 1 cm above the surface of the tray, and are subjected to a standard wetting treatment in a controlled environment chamber. Following 50 mm of spray applied in an initial 2-hr wetting period, the ears are maintained at high relative humidity for 60 hr with the aid of a mister in an air-conditioned chamber held at 20°C ± 1°C. Ears are subsequently dried for 16 hr in a forced air dehydrator at 40°C and the three 10-ear replicates per line are bulked prior to threshing in a Walter and Wintersteiger single plant thresher (Walter and Wintersteiger Gsmbh., Austria Figure 7). The threshed grain is milled in a falling number mill (Falling Number AB, Sweden) to produce a whole-grain flour for determination of falling number or α-amylase activity. This procedure is repeated at 5, 15, and 25 days after eosin ripeness. Typical results for a number of spring wheat cultivars are shown in Table 5. This treatment is also suitable for triticales but is too severe for most barley lines, particularly two-row Australian cultivars. A reduced treatment time would be advisable if falling numbers are being used to assess damage. Once the samples have been dried back to 12% moisture, they can be safely stored for some time without change in α-amylase activity or falling number. In a preliminary study, flour and grain from weather-damaged wheat samples were stored at temperatures ranging from −20 to +40°C. Significant decreases in α-amylase activity, and corresponding increases in falling number, were observed only in the samples stored at 40°C, beginning after about 2 to 3 months of storage. Sprouting tolerance assessment based on falling number and α-amylase activity similar to that described above is a routine selection method in wheat breeding programs in parts of Australia, Canada, Scandinavia, and Europe.

Weilenmann developed a sprouting index (SI) based on falling number values, before and

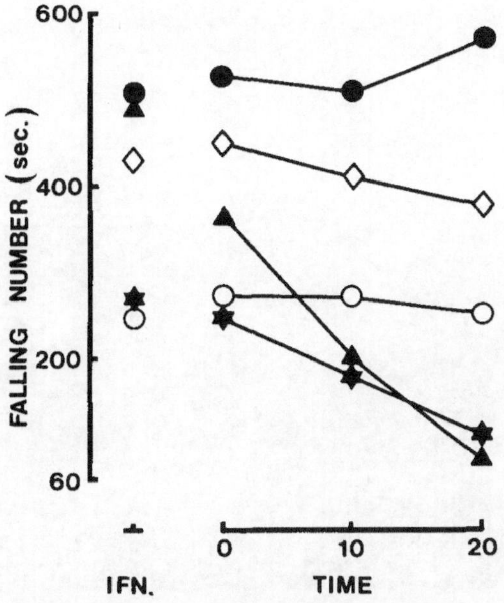

FIGURE 4. Effect of a standard wetting treatment on the falling number of sprouting-tolerant and sprouting-susceptible white wheat cultivars characterized by high or low falling number at maturity in the absence of unfavorable weather. (IFN = falling number at harvest maturity prior to the wetting treatment.) ●, Sprouting tolerant, high IFN — Aus 1408; ◇, moderately sprouting tolerant, high IFN — Suneca; ○, moderately sprouting tolerant, low IFN — Spica; ▲, sprouting susceptible, high IFN — Hartog; ★, sprouting susceptible, low IFN — unnamed breeding line.

after a wetting treatment, measured at intervals during grain maturation and afterripening.[17,40] Using this method, Weilenmann was able to categorize the tolerance of varieties irrespective of differences in maturity and to compare both spring and winter wheats. This method is discussed in greater detail by Derera (Chapter 6).

III. LOCALIZATION AND IDENTIFICATION OF SPROUTING TOLERANCE MECHANISMS

The identification of the mechanism(s) responsible for tolerance to preharvest sprouting, or at least localization of the mechanism to one of the readily defined physiological events in the sprouting process, is an important step towards the efficient utilization of a particular source of tolerance in breeding and cultivar improvement programs. More detailed information on tolerance mechanisms should prove useful in:

1. The development of simple and more specific selection methods,
2. Studies of the effects of environmental factors on tolerance,
3. Attempts to improve the duration and/or the stability of tolerance through a combination of different mechanisms,
4. Genetic studies of the heritability of tolerance, and
5. The eventual identification of the biochemical reactions or factors responsible for tolerance.

FIGURE 5. Hand harvesting of wheat ears.

FIGURE 6. Loading wheat ears onto trays for wetting treatment.

FIGURE 7. Threshing wheat ears in a single plant thresher (Walter and Wintersteiger, Austria).

Table 5
EFFECT OF A STANDARD WETTING TREATMENT ON THE FALLING NUMBER OF RED- AND WHITE-GRAINED WHEATS WITH DIFFERENT LEVELS OF SPROUTING TOLERANCE

			Falling number (sec)			
			15 days		25 days	
Variety	Sprouting tolerance	Grain color	IFN	WD-FN	IFN	WD-FN
RL 4137	Very tolerant	Red	535	523	548	535
Sonora 64A	Susceptible	Red	430	135	441	72
South African 1166	Very tolerant	White	393	431	388	401
Suneca	Moderately tolerant	White	392	349	429	260
SUN 56A	Susceptible	White	351	95	353	65

Note: Samples were grown in field plots at Narrabri in 1984 and treated at 15 and 25 days after eosin maturity. (IFN = falling number prior to treatment, WD-FN = falling number after wetting treatment)

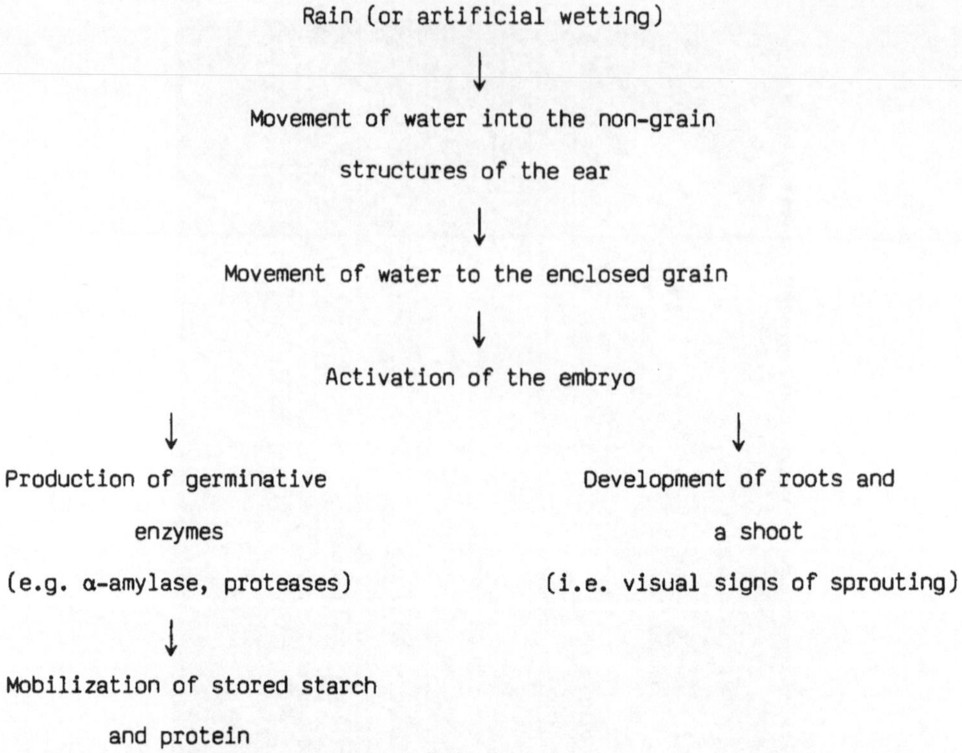

FIGURE 8. Sequence of events in sprouting of cereal ears in response to preharvest rain or artificial wetting treatments.

A. Sequence of Events in Sprouting

Sprouting of cereal grains in the ear prior to harvest can be divided into a number of sequential physiological and biochemical events (Figure 8). Any factor which inhibits or reduces the rate of one or more of the steps prior to the production of hydrolytic enzymes and development of visual signs of embryo germination should give some degree of tolerance. There are consequently a large number of potential mechanisms, although not all would necessarily be compatible with the normal cyclic nature of cereal cropping and end products of the grain. Some mechanisms which ensured the survival of primitive wheats have been inadvertently discarded during the domestication and selection process.[41,42] However, a number of mechanisms still exist in modern wheat cultivars and germplasm collections; some of these are reasonably well understood and characterized, but there are possibly other mechanisms yet to be uncovered. A reduced rate of grain germination at maturity, under conditions normally conducive to embryo activation and germination, commonly referred to as dormancy, has been widely used in agriculture, particularly in red-grained wheat and in barley. Other mechanisms may be useful in combination with dormancy or in situations where inherent dormancy is low, where dormancy is likely to be rapidly dissipated under the weather conditions normally experienced during the harvest period, or where the environmental conditions during grain maturation are unfavorable for the development of dormancy. Several workers have suggested the possible importance of the vegetative structures of the cereal ear in tolerance.[22,23,43,44] Some indication of the effects of the ear, as opposed to factors in the grain, may be obtained from comparisons of germination time for isolated grain and sprouting time in intact ears or spikelets. Where sprouting tolerance is primarily dependent on grain dormancy, then there should be a simple linear relationship between the two measurements. Departures from this relationship, e.g., time for visual sprouting is much longer than would be predicted on the basis of grain germinability, should indicate the presence of factors inhibiting or retarding sprouting located in the vegetative structures of the ear.[22]

B. Estimation of the Rates of Individual Events

In-ear sprouting is a summation of all the events illustrated in Figure 8. The later steps in the sprouting process will obviously be influenced by the preceding steps. As a result, it is often difficult to make direct comparisons of rates of particular events in different cultivars. In attempting to determine the inherent or genetically determined rates of specific steps, the variation in the rates of prior steps in the sequence must be eliminated, or at least reduced in significance, without adversely affecting the step under investigation. In practice, these conditions are difficult to satisfy totally and some compromise must be accepted. For example, it is relatively simple to measure the rate of water movement into isolated grains; however, it is difficult to provide a water source which duplicates the water held by the vegetative structures of the ear. Similarly, for the measurement and comparison of inherent rates of enzyme production, it is essential to reduce the effects of differences in grain germinability.

C. Movement of Water into the Different Anatomical Parts of Cereal Ears

A simple method which has been used for measuring water uptake by wheat ears and grains involves monitoring the moisture content of the relevant tissues at intervals after an artificial wetting treatment. Changes in water content (wet weight − dry weight) of spikes and dissected grain can be used to follow the movement of water into floral organs and grains. King and Richards discussed the relative merits of calculating water uptake on the basis of percentage wet weight or dry weight and concluded that calculations based on wet weight were less likely to emphasize cultivar differences in ear weight.[23] This method can be further refined by dissection of the grain into pericarp, germ tissue, and remaining grain tissues. Since it is the germ tissue water concentration which is important in the initial steps of germination, it is worth while to attempt this measurement. This is reinforced by results discussed by Milthorpe and Moorby which showed a rapid movement of water into the embryo of wheat compared to a much slower increase in endosperm water concentration.[46] For whole grain measurements, differences in germ tissue content would tend to be masked by the relatively large mass of endosperm.

The results discussed by Milthorpe and Moorby pertained to isolated grains, whereas, in terms of understanding field sprouting in cereals, it is clear that the vegetative structures of the spike represent a source of variation in ear water uptake and ear water reservoir size and a potential barrier to the influx of water from the environment.[46] King and Richards reported varietal differences in ear water uptake which they correlated with sprouting tendency.[23] Cultivar differences in water movement into grain in intact ears,[24] and into the germ tissue of grain in ears have also been observed.[47] In this instance the differences could not be readily attributed to differences in spike water content or water uptake by isolated grains. Further investigations in the author's laboratory have shown that rates of water movement into grains in wheat ears were also found to vary with growing environment and to be markedly increased by prior rain or artificial wetting/drying treatments. Rates of water movement did not appear to change with time of storage after harvest or with storage temperature in the range -20 to $+20$°C.[23,47]

Measurements of water uptake by grain in ears are complicated by grain germination and consequently by cultivar differences in germinability. Germination results in rupture of the pericarp/testa over the embryo and rapid growth of the embryo, thus affecting the route of water influx and providing a rapidly expanding sink for water. King and Richards avoided this complication by using after-ripened spikes (6 months storage after harvest),[23] whereas Mares used freshly harvested spikes (no grain germination until at least 36 to 48 hr) and restricted comparisons to water contents below the critical level for germination (35 to 40%, dry weight basis[48,49]).[24] Previous studies indicated that there was no significant difference in initial rates of water uptake by dormant and after-ripened or nondormant grains of a particular wheat cultivar. Water movement into intact wheat spikes has been examined in detail by Bauer and Black[50] and King[33] with considerable attention to wetting rates, wetting intensity, and evaporation.

Water movement into cereal ears consists of three distinct phases:

1. Movement into the vegetative tissues, e.g., glumes
2. Transfer to the surface of the grain
3. Movement into the pericarp, germ, and other tissues of the grain

Genotypic differences have been reported for each of these processes.[22,23,50]

A number of alternate methods have been used to investigate grain water uptake. These include autoradiography following imbibition with tritiated water and the use of colored or fluorescent dyes.[52-55] Another possibility could be the histochemical localization of preexisting enzymes which are activated by the influx of water. These methods have the potential to provide more detailed information on the mode of entry of water into cereal grains. As yet, the autoradiographic methods lack resolution, whereas mature ripe grains appear to be impervious to most dye materials.[54,55] Iodine/potassium iodide solutions have been used with some success to study water movement into the cereal endosperm. However, this method depends on the reaction of the reagent with starch and is, therefore, not suitable for following water movement into germ tissues.

D. Grain Germinability

Grain dormancy is an important factor in the sprouting tolerance of all cereals. Preharvest sprouting or premature germination seriously affects the processing properties of cereal grains. This is particularly serious in wheat and rye where a large portion of annual production is processed for human food. Preharvest sprouting is also detrimental to the production and storage of malting barley and seed grain of wheat, rye, barley, and triticale, since it may adversely affect viability of grain during storage.[56] Prevention of germination is therefore just as important as the prevention of hydrolytic enzyme production.

Grain germination tests are commonly conducted in the presence of excess water (e.g., on moist filter paper or wet sand) at constant temperature (18 to 20°C). Such tests have been criticized on the basis that the conditions are nonphysiological, and indeed Woodbury and Wiebe have shown that germination of wheat grain in the presence of free water is quite different from germination under conditions where water movement outside the grain is restricted.[57] Calculations from data presented by Mares[22] indicate that the water content of the vegetative components of wheat ears following artificial wetting is of the order of 250% dry weight. In standard germination tests, with 4 mℓ of water added to a 9-cm diameter filter paper (dry weight ≈0.5 g), the moisture content of the germination medium may be 800%. Ideally, germination tests should perhaps be conducted at moisture contents closer to that found in wet ears, although there would still be a significant difference in the proportion of the grain surface in contact with the medium. However, while exact duplication of physiological conditions may be important for defining the germination of particular grains in a particular environment, in the present context, where the aim is to identify the relative degree of grain dormancy, then the use of standard conditions under which dormancy is known to be expressed should be sufficient.

Germination may be characterized by measurements of germination rate, the length of the lag phase, the percent germinated grains after a specific period of incubation, or the time required to attain a particular level of germination (e.g., 50%). A number of mathe-

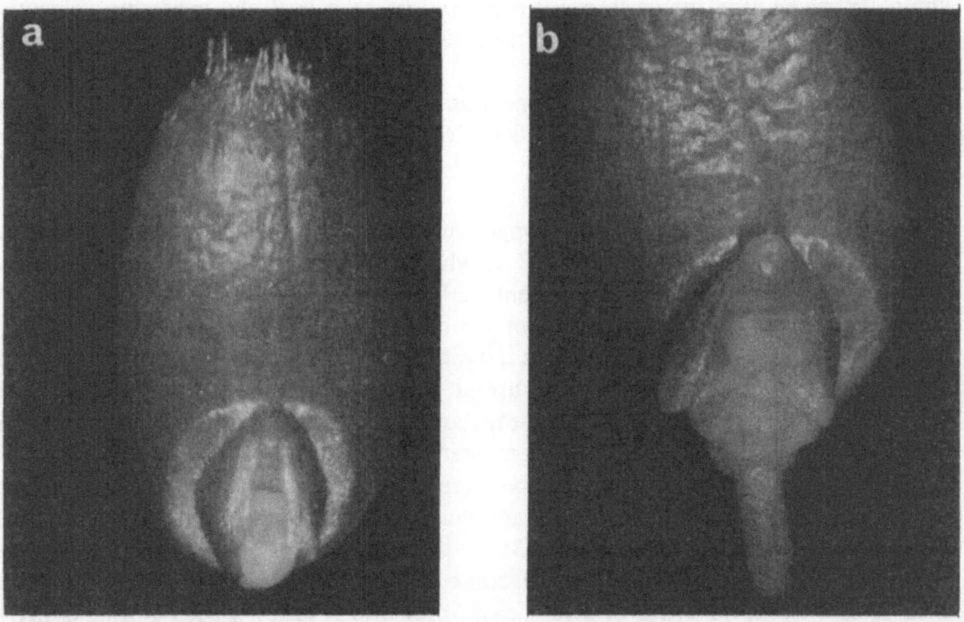

FIGURE 9. Germinated wheat grains. (a) Pericarp over germ ruptured; (b) development of a rudimentary shoot and rootlets.

matical treatments have also been used for comparisons of germination data. These include germination or dormancy index (mean percent germination at 20°C and 12°C with results at 12°C being given double weight),[58] weighted germination percentage, and a range of statistical methods.[10,59,60]

The criterion which is used to decide when germination has occurred may vary depending on the requirements of the investigator. Germination can be judged to have occurred when the pericarp/testa over the embryo is ruptured and the swollen embryo is visible (Figure 9a). Alternatively, the development of a shoot, a seminal root, and two small lateral rootlets may be used (Figure 9b). In the latter case, germination can be readily scored irrespective of the orientation of the grain and, in addition, grains which initiate germination but then fail to proceed further (e.g., because of physical damage or prior sprouting) are not included. During the investigation of germination in a range of wheat genotypes with different levels of dormancy, Mares observed that the time required to progress from one germination stage to the next did not appear to vary significantly between genotypes.[13] In other words, the choice of criterion affected the apparent lag phase but not the slope of the germination curve.

Since germination depends on the stage of maturity,[12,13,60] germination temperature,[12,13,20,31,43,58,61] grain history,[59,62,63] and water and oxygen availability,[57,64] in addition to a genetic dormancy component, it is important to record and standardize test conditions, where possible, and include relevant controls. Gordon has suggested that in addition to the germination test there should be a viability test to eliminate grains which fail to germinate for reasons other than dormancy.[10] Gordon estimated the proportion of viable grains by using a combination of three common dormancy-breaking treatments, viz., cool temperature, dilute potassium nitrate, and alternate light and dark. Nongerminated grains were tested for viability using the triphenyltetrazolium chloride staining technique.[10] Dormancy may also be broken by mechanical rupture or abrasion of the seed coat layers covering the embryo, by the addition of exogenous gibberellic acid, and by the addition of certain other chemicals.[64-66] Embryos taken from dormant cereal grains germinate less readily than embryos taken from nondormant grains. Black et al. concluded that the embryos themselves suffer an internal constraint or inadequacy which is exacerbated by the seed coat and enclosing vegetative tissues (e.g., glumes).[43] Dormancy and germination of cereal grains have been discussed in detail by a number of authors.[43,64,65]

A number of factors affect the measurement or the expression of dormancy in a particular test situation and therefore influence the identification of genetic controls of grain germinability. These include:

1. Temperature: higher germination temperatures (e.g., 30°C) may enhance dormancy, whereas lower temperatures (≤15°C in wheat) enhance the germination of imbibed grain and make it difficult to differentiate between genotypes.[12,13,18,20,31] These temperature effects are only apparent for imbibed grain; for dry grain, increasing temperature up to 40°C simply increases the rate of afterripening.[13] The use of a range of germination temperatures can yield useful information on the relative depth of dormancy and its likely stability in a particular environment. Grain dormancy at harvest ripeness is also controlled by the temperatures which are experienced during grain development. In contrast to the temperature effects noted above, lower temperatures during grain development generally are more favorable for dormancy at harvest ripeness.[59,62,63] Very high temperature (35 to 40°C) during the later stages of grain maturation, which can result in the moisture content of wheat grain declining from 50% to 10 to 12% in the space of 3 to 4 days, may also lead to reduced germinability at harvest ripeness.[47] Presumably this is a result of a gross difference in the rates of grain maturation and grain moisture loss compared to normal.[10]
2. Fungal and/or bacterial contamination: these effects can be largely overcome by a surface sterilization treatment (e.g., sodium hypochlorite, silver nitrate, and Biogram) followed by incubation in a cocktail containing antibacterial and antifungal agents (e.g., 100 mg/mℓ streptomycin plus 100 mg/mℓ mycostatin). Cereal grains, particularly wheat with its deep crease and brush hairs, are often difficult to surface sterilize efficiently with aqueous solutions. Biogram (10 min in 1% vol/vol in water—Gibson Chemicals, Ltd., Australia), a polyphenol preparation containing a surfactant, is very effective and easy to use but must be thoroughly washed off (3 × 5-min washes with water) to avoid residual phytotoxicity.
3. Variability within and between spikes: the errors arising from this phenomenon can be reduced by using an appropriate sample size and random sampling techniques. Black suggested that this phenomenon, heteroblasty, is probably an important facet of sprouting tolerance.[43] Estimates of the degree of variation within seed populations may be as important as the population mean.
4. Water sensitivity: for some cereal grains, semi-dormant barley in particular, excess water may inhibit germination in a proportion of grains.[61,67] A similar condition may also be induced in barley and wheat by the presence of chemicals such as coumarin.[67,68] Water availability also affects germination and interacts with germination temperature.[57]
5. Grain preparation: effects of threshing on grain germinability have already been discussed. In addition, seed weight, seed orientation, and seeding density may also affect germination.[57]

Several investigators have suggested that germination of cereal grain may be partially inhibited by factors residing in the vegetative structures of the ear, in particular the glumes and awns.[25,69] To be of physiological significance these factors must be mobile in order to reach the embryo during the early stages of germination or they must interfere with the influx or transport of oxygen through the seed coat layers. The practice of grinding wheat chaff and spreading it over imbibing grains should be avoided since it enables many chemicals such as toxic products of tissue senescence, which would normally be immobile, to contact the grain surface. Incubation of grain in the presence of intact bracts or parallel observation of germination of dissected grain and grain in intact spikes or spikelets would appear to be more realistic. In addition to germination inhibitors, there is some evidence to suggest that there is genetic variation for sensitivity of the embryo to these chemicals.[70]

E. Production of Germinative Enzymes

During the germination of cereal grains, enzymes such as α-amylase, protease, and cell wall hydrolases are synthesized to mobilize their respective substrates, which are stored in the endosperm, and to provide essential building blocks and nutrients for the growing embryo. These enzymes, endo-enzymes which break covalent bonds within large molecules and render them more susceptible to solubilization and attack by exo-enzymes already present in the mature grain, are active during subsequent processing of cereal flour where they initiate degradation of the grain constituents responsible for processing quality. α-Amylase has attracted the widest attention because of the large and readily measured increases in activity during germination and because of its acknowledged effects on starch paste viscosity and bread baking quality. By contrast, increases in protease activity are less dramatic,[71] are more difficult to measure, and have less marked effects on the commonly used grain and flour quality prediction tests. However, MacGregor has reported that digestion of protein and cell walls in the endosperm of barley during the early stages of germination preceded starch degradation and, indeed, may even be a prerequisite for accessibility of starch to α-amylase.[72] Thus relative rates of production of protease and cell wall hydrolases may determine the extent of movement of α-amylase into the endosperm and have important implications for the salvage of damaged grain via removal of the damaged portion.

Methods for assaying α-amylase and protease are discussed in more detail in Section IV, by Kruger (Chapter 4), and by Mikola.[71]

The quantity produced, the relative rates of synthesis, and the sites of synthesis of these enzymes are all important in determining the response of particular cereal cultivars to rain. In practice the production of enzyme reflects the inherent genetic control of this particular step upon which are superimposed the effects of genotype and environment on all the preceding steps in the sprouting sequence (Figure 8). Gentle hand threshing, which removes the grain from the ear without causing physical damage to the seed coat layers, removes any restrictions or genotypic variation due to the presence of the vegetative structures of the ear. Effects due to variation in germinability are less readily excluded. Culture of varieties under similar environmental conditions or artificial dehydration treatments reduce the environmental variation but not the genetic component of germinability. Dormancy-breaking treatments (see Section II.D of this chapter) are required in this instance. Of these, cool temperature treatment or afterripening are seen as being simple and having less potential to cause unwanted side effects. During afterripening, cereal grains gradually lose dormancy and become more readily germinable. Thus, after a period of time, the genetic differences in grain dormancy decrease in magnitude but do not disappear entirely. Even after 6 months of storage at 12°C, the time required to reach a particular stage in germination still varied by several hours.[47] Mares attempted to further reduce this variation by combining a cool temperature treatment with afterripening.[22,74] Again this treatment reduced, but did not completely abolish, variability in germination between samples. With extended storage, wheat and other cereal grains may lose germination energy, decreasing the apparent initial rate of enzyme production and grain viability. In addition there is some evidence to suggest that low temperature treatment itself may affect enzyme synthesis, e.g., in the gibberellic acid-insensitive (*Rht3*) dwarf wheat 'Tordo'.[75] With each manipulation there is an increased risk of introducing errors, and possibly it would be better to examine cultivars at a number of different dormancy levels and attempt to extrapolate results back to zero dormancy. Using the dormancy-breaking treatments described above, Mares found significant differences in rate, but not the lag phase, of α-amylase production in wheat.[74] The data were consistent with observations made in previous years with the same cultivars.

An alternative, but rather cumbersome, procedure would be to harvest grains at identical morphological stages of germination rather than at specific times. This method would depend on a strict relationship existing between germinative embryo growth and enzyme production. A number of reports indicate that enzyme production can in fact proceed in the absence of embryo growth.[37,38,76-78]

Enzyme synthesis in wheat grain, or other cereal grain, must be examined in relation to the levels of enzyme activity which result in a downgrading in processing quality. Mares and others have reported data relating α-amylase activity in wheat to falling number and amylograph peak viscosity.[74,79,80] These relationships clearly indicate, for example, that very small quantities of enzyme, compared to the total synthetic capacity of the grain, are required to reduce falling number or amylograph peak viscosity to unacceptably low levels. In Australia, the falling number cut off levels of 350, 300, 250, and 200 sec, which have been used in wheat grain segregation at receival, correspond approximately to α-amylase activities of 30 to 40, 50, 75, and 100 to 125 millienzyme units per gram (mEU g^{-1}) compared with sound grain where the activity commonly ranges between 10 and 30 mEU g^{-1}. By comparison, grains which have been incubated at 20°C for 3 days may contain up to 30,000 mEU g^{-1}.[74] Once germinative α-amylase synthesis is initiated there is an immediate effect on falling number or starch paste viscosity, and within a short period maximum allowable levels of activity are exceeded. Thus, in the context of mechanisms of sprouting tolerance, only the initial phase of α-amylase production is relevant. Mares compared two wheat cultivars with quite different initial rates of α-amylase production but similar lag phase. Significant differences in germination time required to produce unacceptable levels of enzyme would only occur when there were only a small number of germinated grains in an otherwise sound sample (1% or less).[74]

Several investigators have reported that during the early stages of germination in cereals, α-amylase is produced by the scutellum, and that only at a later stage is enzyme produced by the aleurone tissue.[72,74,81-83] This information must be considered when designing assay procedures. α-Amylase production by isolated germ tissue explants can apparently account for all the enzyme produced by whole wheat grain over the first 30 hr.[74] Similarly, Gibbons reported that α-amylase produced in barley grains during the first 3 days of germination was derived from the scutellum.[82] Genetic mechanisms, such as gibberellic acid insensitivity, which appear to affect only aleurone α-amylase production are of little relevance to preharvest sprouting in wheat. This information has also proved useful in devising methods for salvaging sound grain from mildly damaged samples.[5] Unfortunately, it is not always possible to predict processing quality of wheat flour from measurements of grain α-amylase and falling number since variable proportions of this enzyme are removed in the bran fraction during milling.[79,84] Unlike the aleurone tissue which appears to store a sizeable proportion of the synthesized enzyme protein, the scutellum appears to release a high proportion of newly synthesized enzyme.[47] The diffusion of the enzyme into the endosperm, and hence its chance of being removed during milling, will presumably depend on the activity of other endosperm-degrading enzymes and the physical nature of the endosperm. This area has not attracted much interest but may well be amenable to genetic manipulation. To the author's knowledge there have been no attempts to screen wheat cultivars for genetic variability in any of these characters.

While selection for low α-amylase-producing genotypes may be worthwhile in wheat, rye, and triticale, it would appear to be counterproductive in barley. Malting, the major process utilizing barley, depends on an even and relatively quick modification of the endosperm by enzymes such as α-amylase and cell wall hydrolases. Unlike dormancy, which declines with afterripening, there is no evidence to indicate that inherent α-amylase synthesis in genetic mutants with low enzyme production rates would change with time.

In addition to enzyme produced by sprouting grains, the processing quality of wheat, rye, and triticale also depends on the level of activity present in mature, apparently sound grain. In general α-amylase (developmental) synthesis reaches a peak during grain development and then declines during maturation to relatively low levels at harvest ripeness (10 to 30 mEU g^{-1} in Australian white wheats). In some wheat genotypes and in most rye and triticale varieties, however, levels of α-amylase activity remain high at maturity even in the absence of unfavorable weather conditions.[22,35,37,38] This is readily demonstrated with an α-amylase or a falling number determination at grain maturity.

F. Susceptibility of Substrates to Attack by Germinative Enzymes

By comparison with the hydrolytic enzymes, the substrate component of the interaction between germinative enzymes and grain constituents has received scant attention. Reports by Moss, Moss and Stiles, and Ringlund have shown that the susceptibility of wheat starch to attack by α-amylase varies considerably with both cultivar and environment.[85-88] Similarly Bushuk and Luckow were able to demonstrate significant differences in the susceptibility of proteins from different wheat cultivars to enzymes produced by sprouted grain.[89] Techniques used by Moss and Stiles and Ringlund involved the collection of large numbers of samples with different levels of weather damage, or blending varying proportions of sound and germinated samples, and the assay of both α-amylase activity and paste viscosity.[86,88] Similar data can be obtained by adding purified α-amylase to sound samples of grain or by using artificial weather treatments to produce a range of damage levels. These results need to be compared with data from field-sprouted samples. Inherent viscosity of wheat starch, and other cereal starches, may also vary with both cultivar and environment.[87,90] This property can be measured on isolated starches or may be estimated by extrapolation to zero α-amylase activity.[86,90] Variation in starch properties and measurement of inherent starch paste viscosity have recently been reviewed by Dengate.[91] Other substrates presumably also show variation in susceptibility to enzymes.

IV. ASSESSMENT OF SPROUTING DAMAGE

A. Industry Requirements

In the event of unfavorable weather conditions affecting a cereal crop prior to harvest, it is imperative that the technology is available to quantify the extent of damage to delivered grain, to enable predictions of the likely effects of the damage on the relevant end use processes, and to ensure that receival standards and segregation guidelines are followed. All sections of the industry require a reliable, objective test for sprouting damage in order to maximize their returns, use grain efficiently, and improve their chances of survival in the current, harsh economic climate. During a rain-affected harvest, grain samples showing a wide range of levels of damage may be delivered. Initially damage may be minimal in relation to effects on the proposed end products, whereas a subsequent small amount of damage may not be obvious from subjective tests but may be readily detected by an objective chemical test. It is this "grey" area where there is the greatest potential for misclassification and comingling of damaged grain with sound stocks. Excessive levels of damage are readily perceived by a well-trained operator and consequently do not present a problem for segregation.

Meredith has reviewed some of the commonly applied sprouting damage test methods which he divides into two groups — "purchase tests", which examine the extent of damage to the grain and the effects of growing, harvesting, storage, and transport; and "quality prediction tests", which inquire into the future and attempt to forecast how flour or grain will perform in particular end product processes.[92] No one test method can realistically be expected to satisfy both requirements or indeed all sections of the grain industry.

Grain growers require an objective, reliable, and reproducible test which enables their grain to be correctly classified or graded on delivery to the local receival depot. Accurate assessment may well mean the difference between making a profit or just breaking even.[93] During the operation of a pilot scheme, based on the falling number method, in northern N.S.W., Australia, objective assessment returned wheat growers an additional A$37 to 45 million which would otherwise have been lost due to inaccuracies in subjective assessment and segregation.[4,93] This savings far outweighed the cost of implementing the scheme. In addition, benefits flowed through the industry due to improved uniformity of wheat grades and hence marketability.

Grain handling or receival authorities are similarly concerned with the accurate classification of incoming loads of grain in order that they can be segregated and binned according to grade. In addition, speed is an important factor if flow through the checking station and unloading are to be maintained. At some receival sites in northern N.S.W., for example, receivals at the peak of harvest may range from 2,000 to 10,000 t/day. Load-by-load testing prior to unloading, such as is practiced in parts of Sweden, would be ideal; however, this is clearly not possible in all cereal-growing regions given the currently available technology. Objective tests applied at receival stations need to be accurate, reproducible, require a minimum of operator technical skill, be robust in situations where conditions may be extreme and variable (e.g., temperature, dust, water quality, and electricity supply), and require a minimum of ancillary equipment and special facilities. Field operation of objective testing requires a well-established central laboratory combined with a sampling system which enables quick and accurate checks of field operations.[4] This is required to ensure the correct application of standards and to check staff and equipment performance and sampling techniques. A number of centers have adopted, or are in the process of adopting, objective testing-based receival and grading systems in conjunction with visual assessment.[2-4,93] These systems vary according to the particular local conditions, receival volume, intake and storage method, and subsequent transport to export site or customer. In Australia, where large volumes of grain are binned at local silos according to grade upon receival, the penalty for misclassification is enormous.

Accurate segregation and specification of grain quality enables marketing organizations to match grain to particular market specifications and to optimize returns for the available grain. In addition, some buying or importing agencies require particular quality specifications such as falling number and amylograph as a condition of sale. Accuracy and reliability help maintain buyer confidence and avoid dramatic fluctuations in quality.

Grain buyers and end users require a reliable and quantitative description of grain quality which enables them to predict the behavior of grain and flour in their particular end product processes. Functional tests (e.g., baking, malting, etc.), which imitate a particular end use product under controlled conditions, are the only tests which can give completely accurate predictions. In practice, predictive test data (e.g., amylograph) and empirical tests (e.g., α-amylase activity, falling number, etc.), provided by the vendor, are used by the grain buyers in the light of previous experience with their particular end products which may vary markedly with respect to their susceptibility to sprouting damage.[8] More and more exporting countries are conducting their own laboratory evaluations of the quality parameters which affect a range of different end products. Different aspects of sprout damage in wheat may vary in their relative effects on particular end products. Thus, α-amylase, which is generally assumed to be the main offender in sprouted grain, is very important in pan bread production; enzyme-mediated darkening reactions and protease activity may be critical for noodle manufacture, whereas grain softening may be a limiting factor in the production of high quality flat bread and chapattis.[8]

In addition there are particular problems for malting and seed grain production which relate to the maintenance of viability of sprouted grain in bulk storage. The link between visibly sprouted grain and storage longevity is well established in barley; however, the relationship between storage longevity and less obvious levels of sprouting damage is not so clear.[56] Currently there are no tests which completely cover this situation. Empirical tests, such as α-amylase and falling number, have been rejected in some instances on theoretical grounds since they measure mean sample damage rather than providing information on the percentage of seeds affected. New fluorescence techniques, discussed later in this chapter, show considerable promise in this area.

B. Visual Assessment of Sprouting

The visual appearance of grain samples is an indicator of the effects of many factors which influence quality, e.g., maturity, hardness, disease, frost damage, degree of contamination or purity, as well as weather damage. Consequently, visual inspection inevitably forms the front line in any classification of incoming cereal grains. In northern N.S.W. it is used as a guide, together with prevailing weather conditions, to determine where and when objective assessment for weather damage should be introduced. In other countries (e.g., Canada) and in barley the visual assessment of grain may be the primary classification criterion at the receival site. Under the Canadian system grain is segregated according to visual appearance on delivery and transported by rail car to the terminal elevators at the seaboard, where it is graded by Grain Commission inspectors prior to bulk binning. During transport, there is time for objective testing to be completed. Misclassification at the site in this system can be rectified at the terminal silo.

Compared to objective tests, visual assessment of sprouting damage suffers from reduced sensitivity and lower discriminating power. In the pilot scheme operated in northern N.S.W., the aim was to segregate within narrow falling number limits (i.e., falling number >350 sec, 350 to 300 sec, and 300 to 250 sec). It is difficult to imagine subjective assessment succeeding in this task.

Visual assessment of the effects of preharvest rain on cereal grain (i.e., the proportion of grains with a ruptured seed coat and perhaps showing evidence of embryo growth) (Figure 10a,b) and its relationship to the degree of sprouting damage is complicated by factors which affect the variation in grain maturity within a crop. These factors include degree of tillering, cultivar, temperature, frost damage, and variation in plant density, moisture stress, and fertility. In addition, the timing of rain with respect to time of harvest ripeness is also important. For a crop which has just reached harvest ripeness (i.e., mean grain moisture equals the moisture level specified for receival) there will be a variable spectrum in terms of degree of grain ripeness and after-ripening and, consequently, considerable variation in grain germinability. Rain at harvest ripeness will usually result in the most mature grains germinating first, and minor levels of weather damage will be characterized by a small proportion of sprouted grains in an otherwise sound sample. For rain falling at some period of time after harvest ripeness (i.e., variation in grain germinability reduced by afterripening) or where rain is accompanied by low temperatures (i.e., expression of grain dormancy depressed) the result can be markedly different. Reduced variation in germinability may be reflected in a large proportion of grains germinating simultaneously and at an equivalent level of weather damage there may be little or no visual evidence of sprouting.

Similarly, there are inherent dangers in a system which classifies grains simply as sprouted or not sprouted. During sprouting, levels of grain α-amylase increase slowly at first and then very rapidly to levels which are many times greater than can be tolerated in most end product processes. Thus 1 or 2% sprouted grains may contain very little enzyme or, at the other extreme, this may be sufficient to downgrade the sample to feed quality. In addition sprouted kernels may be more susceptible to damage during harvesting. The loss of the germ would make it very difficult to assess damage by visual means.

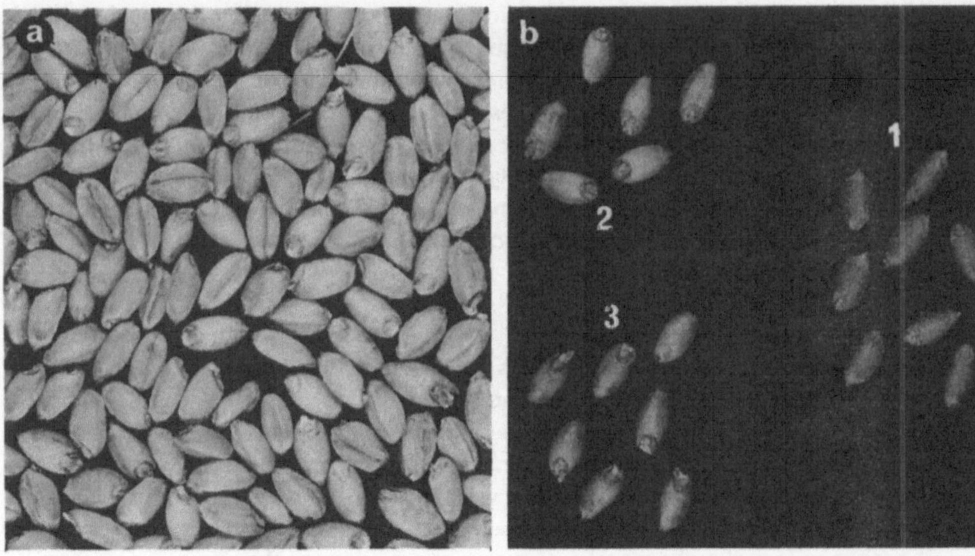

FIGURE 10. Field-sprouted sample of (a) wheat grain and (b) individual kernels (1) showing: (1) no evidence of sprouting, (2) ruptured pericarp, or (3) evidence of embryo growth.

C. Test Weight

Test weight (pounds per bushel or kilograms per hectoliter) is also commonly used as a preliminary indicator of grain quality since many factors such as disease, stress, etc. which affect grain plumpness, and hence bulk density and grain packing characteristics, also indirectly affect milling quality. During wetting and drying cycles, cereal grains swell and expand but may not necessarily return to the original shape or smooth surface texture. Similarly, rupture of the seed coats, embryo growth, and possibly dissolution of some grain constitutents could be expected to decrease test weight. In practice there is a general decrease in test weight with increasing damage; however, the receival agent does not have either a standard curve-relating test weight to weather damage nor a complete history of the particular grain sample. Test weight of a cereal grain sample is dependent on genotype, environmental conditions, and agronomy, and on this is superimposed the effects of sprouting. In addition, the relationship between change in test weight and degree of weather damage may also vary with cultivar and percent of sprouted kernels. Test weight would appear to be of little use in segregation for weather damage except, perhaps, in small, defined geographic locations where receival agents have a sound knowledge of the cultivars grown and some knowledge of test weight prior to unfavorable weather.

D. Falling Number

Hagberg and Perten developed the falling number method as a simple and rapid technique for determining α-amylase activity using flour as the native substrate.[94-96] The falling number system (Figure 11a) utilizes the principle of the rapid gelatinization of a flour suspension with subsequent measurement of the liquifaction of the starch by α-amylase. In the test procedure outlined by the manufacturer (Falling Number AB, Sweden), 300 g of wheat or rye grain (or other cereal) is milled in a Falling Number mill, or similar apparatus, to give a consistent particle size whole meal flour. A representative sample (7.00 ± 0.05 g; sample weight may be adjusted to compensate for variation in moisture content)[102-104] is mixed vigorously with 25mℓ water in a glass viscometer tube and, together with the viscometer stirrer, is placed in the boiling water bath (Figure 11a,b). A motor commences the stirring action after 5 sec and continues for 55 sec. At the end of the mixing cycle the viscometer

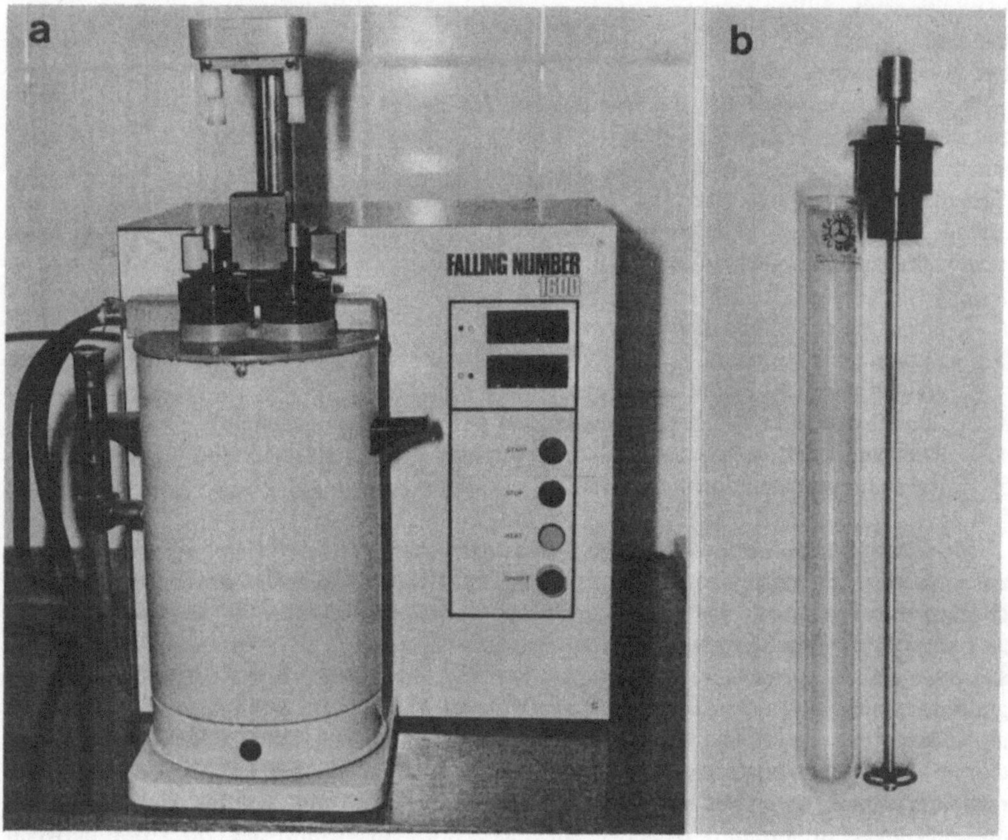

FIGURE 11. (a) Falling number apparatus (Falling Number AB, Sweden); and viscometer tube with viscometer stirrer (b).

stirrer is automatically released from its top position and is free to fall through the gelatinized flour/water suspension. The time required to fall a set distance is recorded in seconds. In the absence of sprouting damage or high initial α-amylase levels the falling numbers are usually high (450 to 550 sec). Increasing levels of sprouting damage and α-amylase result in a decrease in falling number. Since the time is recorded from the moment that the machine is activated, the minimum falling number is 60 sec, and higher levels of α-amylase cannot be measured. Reproducibility between samples and machines is dependent on a uniform rate of heat transfer to the sample; consequently, the viscometer tubes and the stirrer viscometer must be manufactured to rigid specifications. As a result the machine and ancillary glassware are reasonably expensive. Factors which influence the results include sampling procedure, grinding method (method is sensitive to variation in particle size), moisture, stirring technique (largely overcome in automatic models), altitude, water quality, and pH.[97-101] Altitude affects the temperature of the water bath and may cause significant variation above 2500 ft. The temperature of the water bath can be adjusted by the addition of ethylene glycol or glycerol, or alternatively, the falling number values can be mathematically adjusted.[98,101]

The success of this method is reflected in its wide acceptance and application throughout the world and it has been approved by:

- International Association of Cereal Chemistry (IACC; No. 107)
- International Standardization Organization (ISO; No. ISO/DIS 3093)
- American Association of Cereal Chemists (AACC; No. 56-81B)[102-104]

In addition, the falling number value is included in the quality description of wheat sold by the major wheat exporting nations and is used as an official method to segregate wheat and rye in a number of countries.

In the falling number test, the time required for the flour/water slurry to pass through the critical temperature range for α-amylase activity (55 to 80°C, starch gelatinization to enzyme inactivation) is approximately 30 sec and takes place during the initial 60 sec of the test. These conditions are far removed from many industrial processes utilizing wheat or other cereals. Nevertheless, in terms of a "purchase test" or an objective test for grading grain receivals it has a number of advantages. These include:

1. Speed (5 to 10 min, including milling)
2. Reproducibility (error of ± 19 sec for a sample size greater than 250 g,[97] generally quoted as ± 5% of the average)
3. Simple operation which does not require a high degree of technical expertise
4. Robustness and reliability under a wide range of field and laboratory conditions
5. No additional chemicals required

The falling number method measures both enzyme and substrate differences and the results are influenced by genotype and the environmental conditions under which grain samples developed and matured. Both genotype and environment may affect the level of α-amylase at maturity, the rate of enzyme production during sprouting, the inherent viscosity of the substrate and its temperature of gelatinization, the starch content, the proportion of other grain polymers, and the susceptibility of the starch to attack by α-amylase. A measure of the inherent viscosity of the substrate, in the absence of enzyme, may be obtained following a brief acidification treatment to inactivate α-amylase,[100,105] by addition of silver nitrate or mercuric chloride to inhibit α-amylase, or by extrapolation from a series of samples with different levels of sprouting or added enzyme.[86] For sound grain, the correlation between falling number and α-amylase activity is very poor, indicating that substrate differences are significant. However, for a particular sample showing increasing levels of field weathering or artificial weather damage, the correlation is very high. In addition, changes in falling number can be duplicated by the addition of exogenous purified wheat α-amylase to sound flour samples. Changes in falling number which occur during sprouting seem to be dependent almost entirely on changes in α-amylase activity.

A number of research groups have shown a significant interaction between cultivar and the falling number/α-amylase relationship. Finney developed a method which essentially eliminated genetically controlled factors other than α-amylase; however, the procedure necessitated the establishment of standard curves for each wheat cultivar.[106]

Low falling numbers in apparently sound and nonsprouted samples at grain ripeness may be the result of:

1. High α-amylase resulting from either a failure of the developmental α-amylase to be destroyed during normal grain maturation or a genetic or physiological abnormality (common in triticale and also observed in some wheat cultivars) which results in the premature production of germinative alpha-amylase in the absence of embryo germination
2. Low starch paste viscosity conditioned by genotype,[90] genotype-environment interaction, or abnormally rapid grain dessication which leaves starch in a form typical of immature grain

Since its introduction in Europe some 25 years ago there have been numerous studies of the parameters which influence the results of the falling number test and several attempts to improve its performance. Tara and Bains examined the effect of reducing sample size and observed an 18 to 21% reduction in falling number if 5g, instead of 7 g, of flour was used.[107] Reducing sample size also rendered the technique less effective as an index of α-amylase activity. Similarly, Allen and Glennie-Holmes noted that a reduction in sample size exaggerated the effects of extreme growing conditions on falling number.[108] Perten has made two significant modifications to the original method. In the first of these,[109] both cereal and fungal α-amylase activities are measured at 30°C by replacing 50% of the flour with a pregelatinized wheat starch (IACC Method 108).[102] Because fungal α-amylase has low thermostability, it cannot be detected by the standard falling number method. This modification can be used to monitor enzyme supplementation of flour for baking. The second, more recent, modification resulted from an attempt to speed up the analysis time and make the method more practical for load by load testing at receival sites.[110] In this method the test is preformed according to the original specifications except that the stirrer position is recorded after 190 sec using a millimeter scale attached to the mixing head. The report indicates that movement of the viscometer stirrer in this initial 190 sec, although relatively small, is directly related to the final value and could be used as a screening method. For routine purposes the millimeter scale has been replaced by a "falling number predictor flag", i.e., a colured scale with three areas corresponding to predicted falling numbers of >350, >290, and <290 sec. This modification has not been widely tested as yet, but preliminary field tests carried out in northern N.S.W. by the Prime Wheat Association, Ltd. have yielded promising results.

Following a disastrous, weather-damage affected harvest in Australia in 1983/84, the Australian Wheat Board convened a national forum. One of the decisions of this forum was to fund work on the development of a rapid, low cost, objective test which could be used in conjunction with, or instead of, the Falling Number machine and which would allow load-by-load testing of incoming grain. Subsequently, a prototype was developed by a joint CSIRO Wheat Research/Bread Research Institute of Australia team using the viscosity of a gelatinized flour/water paste and a maximum temperature of 93°C (7 g flour and 25 mℓ water as for falling number).[111] "Stirring number" was measured after 2 min by recording the power drawn by a stirring motor that rotated a paddle immersed in the mixture at constant velocity. Special features of the instrument included disposable plastic paddle, aluminum reaction vessel, and an aluminum heating block instead of a boiling water bath. Prototype instruments (Newport Scientific Instruments and Engineering, Australia) were tested in the field in 1985/86 and proved capable of keeping pace with incoming trucks at a major receival site. Correlation of data with falling number was good over a wide range (FN = 62 to 570 sec, r = 0.93) and excellent in the range FN = 62 to 300 sec (r = 0.98). At falling numbers greater than 300 sec, the correlation deteriorated markedly (FN = 300 to 571 sec, r = 0.46 to 0.51). Thus under Australian conditions the correlation began to deteriorate in an area of the falling number spectrum corresponding to the segregation limits for some of the high quality wheat grades (i.e., Australian Prime Hard, FN >350 sec, Northern Hard No. 1 FN >300 sec). The reasons for this sudden breakdown in correlation with falling number is unclear. In the face of this technical problem, the escalating cost of the instrument, and doubts about reproducibility between instruments and operators, the future of the project is uncertain.

The falling number method has withstood the test of time extremely well, a testimony to the simplicity and effectiveness of the original design and subsequent model development, and is still considered to be the best rapid-screening test for hot pasting ability of grain and flour samples.

E. Amylograph

The Brabender amylograph method (Brabender OHG, West Germany; AACC Method No. 22-10) is possibly the most sensitive measure of sprouting damage available and, when

FIGURE 12. Amylograph (Brabender, West Germany): (a) with mixing bowl and (b) mixing head.

used as in investigative procedure, is capable of providing information on inherent paste viscosity, susceptibility of starch to α-amylase, effects of protein and gum components, and damaged starch content, and of giving a reliable prediction of processing properties. After 40 years of experience it is recognized as the industry standard.

The amylograph is a recording viscometer (Figure 12a,b) used primarily to determine the effects of α-amylase on flour paste viscosity as a function of temperature. Starting at around 30°C, the temperature of the flour water slurry (50 to 65 g in 450 mℓ of water) is slowly raised at a rate of 1.5·C/min to 95°C. The time taken for the sample to pass through the critical temperature phase (55°C to 80°C; starch gelatinization to enzyme inactivation) is approximately 15 min. Total test time may vary from 60 to 90 min depending on whether the changes during subsequent cooling are also of concern. The measuring system of the amylography consists of a measuring spring linearly calibrated in cmg (a number of readily interchangeable springs being available, 250 to 1000 cmg) and the measuring range can be extended further by attaching preloaded weights to the measuring system. The characteristics of the pasting curve which are most commonly measured are peak viscosity and peak viscosity temperature.

The amylograph has been discussed in detail by Shuey and Tipples[113] and its application to the assessment of sprouting damage examined by Meredith.[112] Wheat starch pasting properties are significantly influenced by cultivar, while year and location of growth cause a lesser degree of variation. Other factors which affect pasting behavior have been discussed in detail by Dengate.[91] As with the falling number method, inherent starch and flour paste viscosity (i.e., the substrate component of the viscosity measurement) can be determined on isolated starch or following brief acidification treatments of flour to denature α-amylase, respectively.

The large quantity of flour, or starch, required by this method has prompted the development or use of a number of other viscometers, particularly for research on the pasting properties of cereal flour and starch. The Haake Rotovisko uses a much smaller sample size and, in addition, supplies absolute rather than emperical units of viscosity.[91] Other instruments which have been developed include the "Ottawa starch viscometer" and the "minipaster".[114,115] All appear to show good agreement with the Brabender amylograph.

F. Direct Assay of α-Amylase Activity

A wide range of methods have been developed to measure the activity of α-amylase on a standard, often arbitrary, substrate. These methods depend on separation of the enzyme from the grain or flour sample (e.g., extraction with a solution containing 5% NaCl and 0.02% $CaCl_2$ for 5 to 10 min) and provide a good measure of the extent of sprouting damage, but only a rough estimate of likely problems in processing. Most of the methods are time consuming, require some technical expertise and sophisticated equipment, and are generally not suitable for use at grain receival stations.

Older techniques, which generally lacked sensitivity and/or precision, depended on following starch hydrolysis by:

1. Decrease in viscosity
2. Increase in reducing power
3. Decrease in color of iodinated β-limit dextrin

The importance of α-amylase in medical and cereal technology has stimulated the development of techniques which include coupled enzyme reactions enzymatic and automated fluorometric methods, nephelometry, and the hydrolysis of dye-labeled starch.[116-121] As a rule, these direct assays of α-amylase correlate well with other methods such as falling number and amylograph over a wide range of activity or high levels of α-amylase. At low levels of α-amylase, however, other factors, e.g., substrate effects, which are inherent in autolytic methods, become relatively more important.

Two of the recently developed methods which are currently used in the cereal industry are discussed below.

1. Nephelometry

In 1980, Campbell described a new method for the detection of sprout damage in wheat using a nephelometric determination of α-amylase activity.[119] The rate of decrease in turbidity (or light scattering) of a β-limit dextrin solution was followed, using a Perkin-Elmer Model 191 Grain Amylase Analyzer (Perkin-Elmer Corporation, Illinois), and found to increase linearly with the amount of α-amylase. Following several years of collaborative testing, the method was adopted by the American Association of Cereal Chemists (AACC Method No.

22-07), although it does not appear to have found widespread use in the milling and baking industries.[104] The main reason would appear to be the attention to detail required and the degree of laboratory expertise in wet chemistry techniques required.[118]

Further modifications, made to the method by Kruger and Tipples,[120] include:

1. Preparation of own β-limit dextrin substrate form waxy maize starch which is checked daily for possible effects of α-amylase
2. Sensitivity of the instrument increased to its maximum
3. Use of a recorder to follow the decrease in turbidity and allow reaction times to be varied according to the level of enzyme activity
4. Use of fungal α-amylase (5,000 "SKB" units per gram, Calbiochem, Los Angeles, Calif.) as a reference.[120]

This modified method permits the accurate and reproducible determination of low levels of α-amylase in wheat grain and flour and it is currently used by the Canadian Grain Commission to monitor quarterly cargoes of wheat, new lines from plant breeders, and a range of samples submitted to the Research Laboratory.

2. *Dye-Labeled Starch Substrates*

Following a report by Rinderknecht et al.[121] in 1967 of the use of a dye-labeled starch substrate to determine α-amylase activity, several groups have published methods which utilize the principle. These methods, which employ an arbitrary substrate, formed by co-valently linking a soluble dye to a starch derivative, have been compared by Meredith and Pomeranz.[122]

Pharmacia Diagnostics AB (Sweden) developed a cross-linked, partially hydrolyzed potato starch substrate with a 3-D lattice configuration, which is insoluble in water. The substrate is labeled with Cibachron Blue and appears to be completely resistant to attack by β-amylase. The substrate is marketed in powder form or as tablets (Phadebas tablets [Pharmacia Diagnostics AB] which include sodium phosphate buffer (pH 7.0), stabilizer, and salt, in addition to 45 mg dry blue starch polymer.[123] Each batch of tablets is accompanied by a standard calibration curve prepared using human α-amylases. Barnes and Blakeney used this substrate for the determination of cereal α-amylase and examined effects of temperature, starch damage, pH, and calcium ion concentration and the relationship to falling number and amylograph viscosity.[123]

In the method described by Barnes and Blakeney, ground wheat (1 g) or flour (5 g) is extracted for 5 min with 20 mℓ of (1) 0.05 *M* maleate buffer (pH 6.0) containing calcium chloride (0.2 g/ℓ) or (2) distilled water containing sodium chloride (5 g/ℓ) and calcium chloride (0.2 g/ℓ) and then centrifuged. The 5 mℓ of extract is incubated for 15 min at 50°C with one Phadebas tablet with shaking every 5 min. The reaction is stopped by addition of 1 mℓ 0.5 *M* NaOH; the reaction mixture is diluted to 10 mℓ, filtered, and the absorbance of the colored solution is measured at 620 nm. Barnes and Blakeney calibrated the tablets against purified wheat α-amylase and derived conversion factors which could be used, in conjunction with the calibration chart supplied by the manufactuer, to calculate millienzyme units of wheat α-amylase per 10 mℓ of colored solution [(U/ℓ) from chart × 0.025 (maleate buffer) or × 0.04 (tablet alone)].[123] This assay procedure requires in excess of 30 min to complete and is clearly not compatible with requirements of grain receival. Nevertheless, the method is very useful in routine laboratory and research applications. A semiautomated, instrumental version of this method, the Deltron Grain Tester has been designed and manufactured by Altech Associates Pty., Ltd., Australia. This instrument can perform two tests simultaneously but must be supplemented with a spectrophotometer for quantitative determination of α-amylase.

In the U.S. potato amylase dyed with Cibachron F 3GA (Hoffman La Roche, U.S.) was

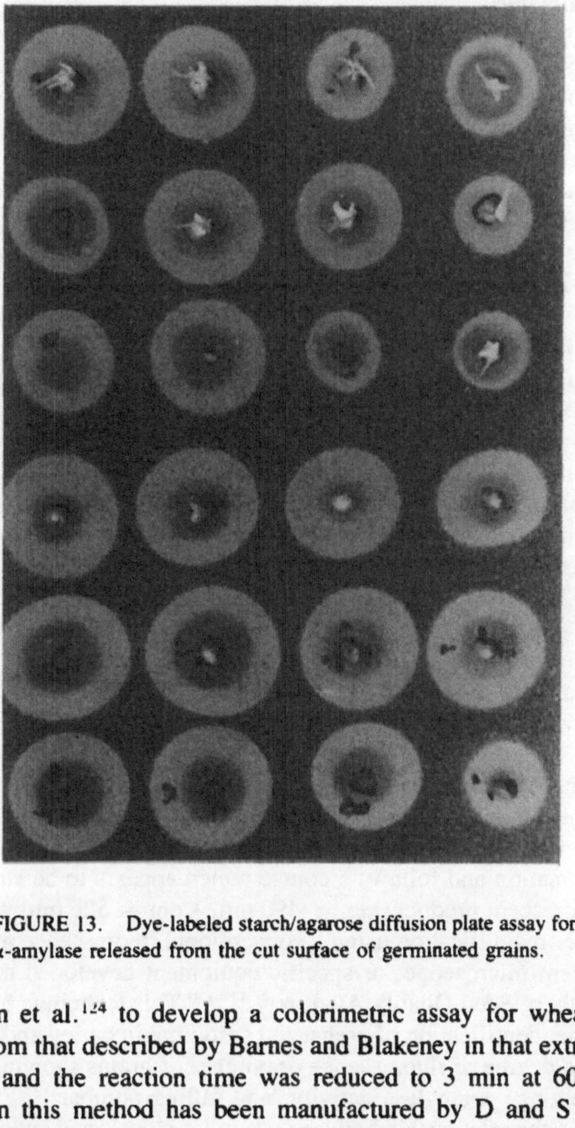

FIGURE 13. Dye-labeled starch/agarose diffusion plate assay for α-amylase released from the cut surface of germinated grains.

used by Mathewson et al.[124] to develop a colorimetric assay for wheat α-amylase. This procedure differs from that described by Barnes and Blakeney in that extraction and reaction were simultaneous and the reaction time was reduced to 3 min at 60°C. A commercial instrument based on this method has been manufactured by D and S Instruments, Ltd., Pullman, Wash., U.S.

Dye-labeled starch substrate has also been incorporated into agarose gel slabs and used to assay cereal α-amylase activity by radial diffusion, diffusion zone diameters being proportional to the logarithm of α-amylase activity. In the procedure described by Hejgaard and Gibbons, agarose (150 mg) is dissolved in boiling 50 m M phosphate buffer, pH 6.9 (15 mℓ) containing 20 m M $CaCl_2$ and 40 mg blue starch powder (Pharmacia Norden AB, Uppsalla, Sweden) and poured onto a 10 × 10-cm glass plate to a depth of 1.5 mm.[125] Extracts (25 per plate) are placed in sample wells cut in the solidified gel and incubated at 25°C overnight for 2 to 4 hr at 50°C. The reaction is stopped by immersing the plate in 5% acetic acid for 5 min; then the gels are washed, pressed under a wad of filter paper (uniform pressure of 10 g/cm^2), and dried to give a permanent record. The method uses the equivalent of one Phadebas tablet for 25 assays, represents a considerable saving in cost and labor, and is particularly suited to screening large numbers of samples.

The method can also be used to measure α-amylase diffusing from cut grain surfaces (Figure 13), and to monitor column chromatography eluates. The blue starch powder can be replaced by a Phadebas tablet; however, the presence of tableting agents makes it more difficult to achieve a uniform substrate distribution and impossible to dry the gels.[125]

G. Use of Other Enzymes

Many enzymes, in addition to α-amylase, increase during the early stages of grain germination and potentially could be used to measure or quantitate sprouting damage in cereal grains. Cell wall polysaccharide-degrading enzymes and endo-proteases appear to be important for providing access of α-amylase to the surface of starch granules in cereal endosperm,[72] in addition to providing metabolites for embryo growth, and may well be good indicators of incipient sprouting damage. Unfortunately, current assay techniques for these enzymes have neither the sensitivity or the reliability required for a field screening test. The role of these enzymes, together with lipases, enzymes catalyzing darkening reactions, etc., in cereal product processes and quality has been largely neglected.

H. Fluorescent Methods

In this section a range of relatively new and very sensitive techniques, which use fluorescent substrates or fluorescent products to follow enzymic reactions or attach specific, fluorescence-labeled antibodies to specific grain constituents, are discussed.

Gibbons used these techniques in a study of the sequential determination of α-amylase transport and cell wall breakdown in germinating barley seeds.[82,126] Monospecific antibodies to purified α-II-germination-specific α-amylase were labeled with fluorescein and used to tag enzyme protein excreted from the scutellum and later the aleurone tissue. Simultaneously, Calcofluor White M2R New, which forms a fluorescent complex with endosperm cell wall β(1 → 3, 1 → 3) glucans, was used to study cell wall dissolution. Unfortunately, the calcofluor stain is not as useful in other cereal grains such as wheat where β-glucans are only a minor cell wall component. In barley, at least, the technique allowed the degree of modification of individual grains to be visualized and correlated with events in the malting process.

Further work at the Carlsberg Research Centre in Copenhagen, Denmark has resulted in the application of the fluorochrome, fluorescein dibutyrate (FDB), to the study and monitoring of germination in cereal grains. The action of a lipase, which is produced during the early stages of germination and follows a course which appears to be similar to α-amylase, releases a highly fluorescent product ($\lambda ex = 490$ nm, $\lambda\ em = 520$ nm) which can be readily seen under low power magnification using a conventional fluorescence microscope, a Carlsberg macro-fluorescent microscope, or specific equipment developed at the Carlsberg Research Centre (Carlsberg Seed Quality Analyzer).[127-130] This technique has been assessed as a screening tool in the identification of preharvest sprouting in barley and wheat and attempts made to correlate the degree of fluorescence or number of grains showing fluorescence with degree of field sprouting, α-amylase activity, and falling number.[127-130] Large differences have been observed in the relationship between α-amylase levels and endosperm modification in different wheat and barley cultivars.[129] The method involves several sequential preparative steps, including the staining of the cut surface of grains by immersion in 1 m*M* fluorescein dibutyrate (Serva Feinbiochemica, Heidelburg, West Germany) in 80% (v/v) ethanol for 2 min, followed by a subjective estimate of extent of fluorescence; it is consequently difficult to envisage the method being applied at receival sites for the determination of sprouting damage.

The method does have exciting prospects, however, in other areas of the cereal industry. Because individual grains which are pregerminated can be quickly and quantitatively determined (compared with α-amylase or falling number methods, which give only a mean sample level of damage) the FDB method is particularly useful in assessing effects on malting quality, in following the malting process, and in predicting storage longevity of sprouted grain samples. Examination of barley samples with 0 to 40% pregerminated grains showed a significant inverse relationship between pregermination percentage and viability after 12 months storage.[127] Pregermination as low as 2 to 3% resulted in significant reductions in viable barley grains after storage.

Another application of the FDB method involves the salvage of sound grain from weather-damaged samples. Separation procedures based on a pneumatic gravity table have been fine-tuned to remove damaged grains using the FDB method.[6]

I. Relationships between Different Methods

In view of the large number of different methods which can be used to identify and quantify sprouting damage in cereal grains, it is important to understand how the results of one test correlate with data from another method. Subjective tests such as visual scores of sprouted grains (i.e., grains classified as sprouted or unsprouted) or objective tests such as test weight, which are affected by a wide range of factors other than sprouting damage, cannot realistically be expected to show a high correlation with more specific objective tests such as α-amylase activity or falling number. Similarly, determinations of α-amylase in grain or flour extracts using a standard substrate cannot be expected to show a strong correlation with autolytic methods at low levels of α-amylase where substrate effects become significant. Attempts to extrapolate from one method to another will be subject to errors resulting from inherent differences in the way the different methods measure sprouting damage even though each method, in its own right, may be a good measure of damage.

Both falling number and amylograph viscosity show a curvilinear relationship with α-amylase activity.[79] These relationships can be transformed to linear functions by first converting falling number to liquifaction numbers(LN = 6000/FN-50) and amylograph viscosity to mobility (1[one]/peak viscosity).[80]

α-Amylase activity is strongly correlated with both falling number (r = 0.86 to 0.975) and amylograph viscosity (r = 0.86 to 0.91) over a wide range of values.[79,96,123] Over the much narrower range that is relevant to the segregation of weather-damaged wheat in practice, however, these correlations are much weaker. Calculations from data reported by Barnes and Blakeney[123] gave correlation coefficients of r=0.76 and r=0.61, respectively, for the relationships between α-amylase and falling number (FN = 200 to 400 sec) and amylograph viscosity (viscosity = 200 to 750). At a particular level of α-amylase activity, the falling number may vary over a range of approximately 150 sec. While α-amylase activity may be a good measure of weather damage, it cannot be used, with any confidence, to predict falling number accurately.

The relationship between falling number and amylograph peak viscosity is subject to considerable variation due to inherent differences in the method of measuring viscosity and sample preparation. In the amylograph method, the heating rate is programmed at a much slower rate and as a consequence thė rate of gelatinization of starch and the rate of heat inactivation of enzymes will be different than for the falling number. Corr and Spillane observed differences in falling number in flour samples with similar α-amylase activities and concluded that the differences were a function of starch susceptibility caused by the initial gelatinization temperature.[131] Due to the rapid rise in temperature in the falling number method, hydrolysis time is a limiting factor, and any small differences in early gelatinization behavior are reflected in the results. Such differences would have little influence on the amylograph viscosity levels.

Another important difference between the falling number method and the amylograph concerns the method of sample preparation. Falling numbers are usually determined on whole grain milled to a meal of consistent particle size. By constrast, amylographs are carried out on milled flour whose properties may differ markedly, depending on extraction rate, type of mill, and degree of flour refinement and bran contamination. Kruger and Tipples compared 170 samples of Canadian wheat and found a good correlation between liquifaction number and mobility (r = 0.91).[79] Unfortunately, there were a number of anomalous data points which appeared to result from variation in the proportion of the enzyme which ended up in the flour. Kruger and Tipples subsequently examined this aspect more closely and in 167 samples of wheat they observed a wide variation in the proportion of grain enzyme which was retained in the flour (10 to 80%, mean value of 40 to 50%).[79] This variation would be expected to have a profound effect on the relationship between falling number (whole meal) and amylograph viscosity (milled flour).

J. Sprouting Damage Prediction

A scheme or model which could be used to predict the relative level of sprouting tolerance at maturity in a cereal crop would have obvious advantages for farmers and grain handling authorities. Farmers might be able to adjust their harvesting schedules to reduce the rate of sprouting, while grain handling authorities might be able to make more effective use of their objective testing facilities. Several attempts have been made to develop such a predictive model for cereals based on environmental conditions during grain development. Belderok reported that there was an inverse relationship between the temperature sum (product of days and temperature in excess of 12.5°C) during the dough ripe stage of grain development (45 to 22% moisture) and the length of postharvest dormancy.[132] Other research groups have had less success with this model.[133-135] Olsson and Mattsson observed that the influence of accumulated temperature during the dough stage on dormancy varied from year to year and with location.[134] These workers suggested that other climatic factors were also important. Grahl and Schrödter tested Belderok's model using winter and summer wheat cultivars grown in different regions of West Germany.[133] They found no relation between the sum of the daily mean temperature above 12.5°C during the dough stage and dormancy in either maritime or continental environments. Reiner and Loch identified two temperature-sensitive periods in the development of barley grain.[136] During the first period (12 to 16 days after ear emergence), low temperatures reduced the level of dormancy at maturity. In contrast, during the second sensitive period (dough ripe stage, 30 to 41 days after ear emergence) high temperatures reduced, and low temperatures increased, the level of dormancy achieved at maturity. In a more recent study, Nielson showed that precipitation after morphological maturity (maximum grain weight) had a significant influence on sprouting tolerance.[137] Little follow up work appears to have been carried out in this area.

Although not a true predictive scheme, the falling number prognosis scheme in operation in the Scania region of Sweden is worthy of note. During grain development and maturation, samples are regularly collected from crops throughout the region for determination of ripeness, falling number, and moisture. This information is used to produce a daily bulletin of trends in falling number and may also give advice with respect to harvesting.[3] Such a scheme would appear to be ideally suited to cereal-growing regions with maritime climates characterized by low temperatures and frequent, but light, precipitation during the grain maturation and harvest periods. In other areas or countries, rain at or near harvest ripeness is either unseasonal and irregular (perhaps causing preharvest sprouting in 1 out of 5 or 10 years for example) or of short duration but very heavy (summer rainfall areas such as northern N.S.W. and Queensland, Australia). In these situations there is no gradual decline in falling number but a sudden fall resulting from an isolated rainy period and any forecast of possible damage is impossible.

V. DISCUSSION

Older, traditional cereal-processing methods and small manufacturing units are increasingly being replaced by larger, centralized processing facilities throughout the world. With the influx of modern, high technology, stricter quality controls, computer-controlled and automated production lines, and a requirement for greater product uniformity and quality, there is a continuing need to improve and develop new techniques for detecting and quantifying the effects of sprouting on cereal grain constituents and their functional properties. In some instances, further progress will be dependent on a better understanding of the basic biochemistry and physiology of sprouting and the exact role of grain constituents in a wide and varied range of end product processes.

Cultivar improvement through breeding and selection for sprouting tolerance is a lengthy and expensive exercise, partly because of the difficulties in handling a character which is influenced by a large number of environmental and agronomic, as well as genotypic, factors. Elucidation of the underlying genetically controlled biochemical or physiological events which determine tolerance may eventually lead to the development of screening tests, which are specific for particular tolerance mechanisms, based on genotypic characters. The pathway of water and gas movement into grains in intact ears is poorly understood and, despite a considerable amount of research, grain dormancy is still an elusive phenomenon. Some aspects, by comparison, are relatively well understood. For example we now have a good working knowledge of α-amylase synthesis, control, and secretion in germinating cereal grains, of the action of this enzyme on starch granules and gelatinized starch, and of its effects on a number of end products. Practical applications of this knowledge are widespread in breeding, quality assessment, utilization, and salvaging operations related to preharvest sprouting.

To date most work has focused on the readily measured enzyme α-amylase. The role of other enzymes and indeed, other constituents of sprouted grains in end product processing, has only recently begun to attract attention. In the future, tests for some of these constituents may need to be developed to assess more accurately the effects of sprouting on processing properties. Increasingly, customers around the world are showing a higher level of expertise in quality analysis and are demanding more extensive and more precise quality information prior to sale.

A rapid, cheap, and reliable test of preharvest sprouting damage which can be applied in the field on a load-by-load basis is urgently required in some countries. The falling number method is tantalizingly close to at least some of these requirements and possibly recent modifications may prove successful. Any new test must be directly and accurately correlated with the currently accepted international measure, falling number, at low levels of α-amylase and the higher end of the falling number scale (200 to >350 sec) where segregation is applied in practice. It is not sufficient to show a good correlation over a wide range of α-amylase activities or falling number values. In the short term, these restrictions probably limit new tests to autolytic methods similar to falling number or the amylograph. For a new test, which uses a different principle or measures a different component of sprouting damage, to be approved, significant advantages in accuracy and quality prediction would have to be demonstrated. The introduction of a new standard would require considerable changes throughout the industry and a reeducation of consumers who have established a relationship between the current sprouting damage test data and their particular end products.

Another area which deserves attention is the influence of incipient sprouting on the viability of cereal grain during storage.

The cereal industry encompasses a wide range of people, organizations, and technological processes. Changes made at any stage of the system must be weighed in relation to their effects on end products and consumer acceptability. Thus, there is little point to developing a new cultivar which maintains a high falling number, despite rain prior to harvest, if some other constituent results in a product of low consumer appeal.

VI. ACKNOWLEDGMENTS

The author would like to thank Mr. E. Czarnecki, Dr. L. Song, and Dr. R. Henry, who provided details of their sprouting tolerance screening and testing programs. Some of the work reported was funded by grants generously provided by the Wheat Industry Research Council of Australia, the Wheat Industry Research Committee of N.S.W., and the Prime Wheat Association, Narrabri.

REFERENCES

1a. Proc. First International Symposium on Pre-Harvest Sprouting in Cereals, *Cereal Res. Commun.*, 4, 1976, pp. 81—273.
1b. Proc. Second International Symposium on Pre-Harvest Sprouting in Cereals, *Cereal Res. Commun.*, 8, 1980, pp. 15—302.
1c. *Proc. Third International Symposium on Pre-Harvest Sprouting in Cereals*, Kruger, J. E. and LaBerge, D. E., Eds., Westview Press, Boulder, Colo., 1983.
1d. *Proc. Fourth International Symposium on Pre-Harvest Sprouting in Cereals*, Mares, D. J., Ed., Westview Press, Boulder, Colo., 1987.
1e. **Briggle, L. W.**, Pre-harvest sprout damage in wheat in the U.S., *Cereal Res. Commun.*, 8, 245, 1980.
2. **Kruger, J. E.**, The Canadian grading system and implications for monitoring of sprout damage, in Reference 1d, p.543.
3. **Larsson, S.**, Falling number prognoses on rye and wheat, in Reference 1d, p.550.
4. **Keene, T.**, Assessment of weather damage under field conditions in northern N.S.W., in Reference 1d, p.555.
5. **Henry, R. J., Martin, D. J., and Blakeney, A. B.**, Pearling and milling as techniques for the improvement of the quality of sprouted wheat, in Reference 1d, p.207.
6. **Munck, L.**, The control of pre-harvest sprouting in cereals for seed, malting and milling, in Reference 1d, p.176.
7. **Weipert, D.**, Processing of grain according to alpha-amylase activity and preparing optimal blends from sound and damaged wheat, in Reference 1d, p.197.
8. **Orth, R. A. and Moss, H. J.**, The sensitivity of various products to sprouted wheat, in Reference 1d, p.167.
9. **Czarnecki, E.**, personal communication, 1986.
10. **Gordon, I. L.**, Germinability, dormancy and grain development, *Cereal Res. Commun.*, 8, 115, 1980.
11. **Gordon, I. L., Balaam, L. N., and Derera, N. F.**, Selection against sprouting damage in wheat. II. Harvest ripeness, grain maturity and germinability, *Aust. J. Agric. Res.*, 8, 115, 1980.
12. **Gosling, P. G., Butler, R. A., Black, M., and Chapman, J. M.**, The onset of germination ability in developing wheat, *J. Exp. Bot.*, 32, 621, 1981.
13. **Mares, D. J.**, Temperature dependence of germinability of wheat (*Triticum aestivum* L.) grain in relation to pre-harvest sprouting, *Aust. J. Agric. Res.*, 35, 115, 1984.
14. **Copeland, P. J. and Crookston, R. K.**, Visible indicators of physiological maturity in barley, *Crop Sci.*, 25, 843, 1985.
15. **Hanft, J. M. and Wyck, R. D.**, Visual indicators of physiological maturity of hard red spring wheat, *Crop Sci.*, 22, 584, 1982.
16. **Clarke, J. M.**, Time of physiological maturity and post-physiological maturity drying rates in wheat, *Crop Sci.*, 23, 1203, 1983.
17. **Weilenmann, F.**, A selection method to test the sprouting resistance in wheat, *Cereal Res. Commun.*, 4, 267, 1976.
18. **Salmon, D. F. and Helm, J. H.**, Pre-harvest and post-harvest dormancy in spring triticale, *Agron. J.*, 77, 649, 1985.
19. **Clarke, J. M.**, Wheat weathering damage in windrower/combine vs. direct combine harvesting systems, in Reference 1c, p.287.
20. **Noll, J. S. and Czarnecki, E.**, Methods for extending the testing period for harvest-time dormancy in wheat, *Cereal Res. Commun.*, 8, 233, 1979.
21. **Mares, D. J.**, Preservation of dormancy in freshly harvested wheat grain , *Aust. J. Agric. Res.*, 34, 33, 1983.
22. **Mares, D. J.**, Pre-harvest sprouting tolerance in white grained wheats, Reference 1d, p.64.
23. **King, R. W. and Richards, R. A.**, Water uptake in relation to pre-harvest sprouting damage in wheat: ear characteristics, *Aust. J. Agric. Res.*, 35, 327, 1984.
24. **Mares, D. J.**, Investigation of the pre-harvest sprouting damage resistance mechanisms in some Australian white wheats, in Reference 1c, p. 59.
25. **Derera, N. F. and Bhatt, G. M.**, Germination inhibition of the bracts in relation to pre-harvest sprouting tolerance in wheat, *Cereal Res. Commun.*, 8, 199, 1980.
26. **McCrate, A. J.**, Pre-harvest Sprouting and Seed Dormancy in Hard Red and Hard White Winter Wheat (*Triticum aestivum* L.), Ph.D. thesis, Kansas State University, Manhattan, 1981.
27. **Noll, J. S., Dycke, P. L., and Czarnecki, E.**, Expression of RL 4137 type of dormancy in F_1 seeds of reciprocal crosses in common wheat. *Can. J. Plant Sci.*, 62, 345, 1982.
28. **Song, L.**, personal communication, 1986.
29. **Larsson, S.**, Selection for seed dormancy using a simple germination test, see Reference 2, p.400.

30. **Reitan, L.**, Early generation mass selection for seed dormancy in barley, in Reference 1c, p.244.
31. **George, D. W.**, High temperature dormancy in wheat (*Triticum aestivum* L.), *Crop Sci.*, 7, 249, 1967.
32. **McMaster, G. J. and Derera, N. F.**, Methodology and sample preparation when screening for sprouting damage in cereals, *Cereal Res. Commun.*, 4, 251, 1976.
33. **King, R. W.**, Ear and grain wetting and pre-harvest sprouting, in Reference 1d, p.327.
34. **Henry, R. J. and McLean, B. T.**, Relative rates of sprouting, alpha-amylase production and endosperm breakdown during intact head wetting at different temperatures, in Reference 1d, p.131.
35. **Mares, D. J.**, Comparisons of tolerance to pre-harvest sprouting within *Triticum* species, triticale and barley, in Reference 1d, p.148.
36. **Gordon, I. L.**, Heritability of grain development traits associated with sprouting damage in wheat inbreds, *Cereal Res. Commun.*, 8, 185, 1980.
37. **Gale, M. D., Flintham, J. E., and Arthur, E. D.**, Alpha-amylase production in the late stages of grain development — An early sprouting damage risk period, in Reference 1c, p.29.
38. **Gale, M. D., Salter, A. M., and Lenton, J. R.**, The induction of germination alpha-amylase during wheat grain development in unfavourable weather conditions, in Reference 1d, p.273.
39. **Oettler, G.**, Variation of falling number in primary triticale and their wheat and rye parents, in Reference 1d, p.123.
40. **Weilenmann, F.**, Ten years experience with using sprouting index in wheat breeding, in Reference 1d, p.93.
41. **Mac Key, J.**, Seed dormancy in nature and agriculture, *Cereal Res. Commun.*, 48, 83, 1976.
42. **Mac Key, J.**, Dormancy in the wild relatives of modern cereals, in Reference 1d, p.414.
43. **Black, M., Butler, J., and Hughes, M.**, Control and development of dormancy, in Reference 1d, p.379.
44. **Reitan, L.**, Breeding and selection for pre-harvest sprouting tolerance in barley, in Reference 1d, p.75.
45. **Gordon, I. L., Derera, N. F., and Balaam, L. N.**, Selection against sprouting damage in wheat. I. Germination of unthreshed grain, with a standard wetting procedure, *Aust. J. Agric. Res.*, 28, 583, 1977.
46. **Milthorpe, F. L. and Moorby, J.**, *An Introduction to Crop Physiology*, 2nd ed., Cambridge University Press, Cambridge, 1979.
47. **Mares, D. J.**, unpublished data, 1986.
48. **Owen, P. C.**, The relation of germination of wheat to water potential, *J. Exp. Bot.*, 3, 188, 1952.
49. **Lush, W. M., Groves, R. H., and Kaya, P. E.**, Presowing hydration-dehydration treatments in relation to seed germination and early seedling growth of wheat and ryegrass, *Aust. J. Plant Physiol.*, 8, 409, 1981.
50. **Bauer, A. and Black, A. L.**, Sprouting in intact mature spikes of hard red spring wheat, *Agron. J.*, 75, 1016, 1983.
51. **King, R. W.**, Water uptake in relation to pre-harvest sprouting damage in wheat: grain characteristics, *Aust. J. Agric. Res.*, 35, 337, 1984.
52. **Stenvert, N. L. and Kingswood, K.**, An autoradiographic demonstration of the penetration of water into wheat during tempering, *Cereal Chem.*, 53, 141, 1976.
53. **Jackson, G. M. and Varriano-Marston, E.**, A simple autoradiographic technique for studying diffusion of water into seeds, *Plant Physiol.*, 65, 1229, 1980.
54. **Briggs, D. E. and MacDonald, J.**, The permeability of the surface layers of cereal grains and implications for tests of abrasion in barley, *J. Inst. Brew.*, 89, 324, 1983.
55. **Cook, H. and Oparka, K. J.**, Movement of fluorescein into isolated caryopses of wheat and barley, *Plant, Cell Environ.*, 6, 239, 1983.
56. **Moor, T.**, Problems encountered in storage of sprout damaged barley, in Reference 1d, p.212.
57. **Woodbury, W. and Wiebe, T. J.**, Temperature-moisture interactions as a factor in germination and dormancy, in Reference 1c, p.43.
58. **Strand, E.**, Effect of temperature and rainfall on seed dormancy of small grain cultivars, in Reference 1c, p.260.
59. **Reddy, L. V., Metzger, R. J., and Ching, T. M.**, Effects of temperature on seed dormancy of wheat, *Crop Sci.*, 25, 455, 1985.
60. **Scott, S. J., Jones, R. A., and Williams, W. A.**, Review of data analysis methods for seed germination, *Crop Sci.*, 24, 1192, 1984.
61. **Strand, E.**, A seed dormancy index for selection of cultivars of cereals resistant to pre-harvest sprouting, *Cereal Res. Commun.*, 8, 219, 1980.
62. **Belderok, B. and Haberkotté, A.**, Induction of secondary dormancy in wheat crops by low temperature and high relative humidities in the field, *Cereal Res. Commun.*, 8, 167, 1980.
63. **Takahashi, N.**, Effect of environmental factors during seed formation on pre-harvest sprouting, *Cereal Res. Commun.*, 8, 175, 1980.
64. **Taylorson, R. B. and Hendricks, S. B.**, Dormancy in seeds, *Annu. Rev. Plant Physiol.*, 28, 331, 1977.
65. **Roberts, E. H. and Smith, R. D.**, Dormancy and the pentose phosphate pathway, in *The Physiology and Biochemistry of Seed Dormancy and Germination*, Khan, A. A., Ed., North-Holland, Amsterdam, 1977, 385.

66. **Cairns, A. L. P. and de Villiers, O. T.**, Breaking of seed dormancy by treatment with ammonia, in Reference 1d, p.393.
67. **Briggs, D. E.**, *Barley*, Chapman and Hall, London, 1978.
68. **Woodbury, W. and Wiebe, T. J.**, A possible role for the pericarp in control of germination and dormancy of wheat, in Reference 1c, p.51.
69. **Derera, N. F., Bhatt, G. M. and McMaster, G. J.**, On the problem of pre-harvest sprouting of wheat, *Euphytica*, 26, 299, 1977.
70. **Stoy, V. and Olsen, O.-A.**, Inheritance of a factor affecting the response to germination inhibitors in excised wheat embryos, *Cereal Res. Commun.*, 8, 203, 1979.
71. **Mikola, J.**, Proteinases and peptidases in germinating cereal grains, in Reference 1d, p.463.
72. **MacGregor, A. W.**, Cereal endosperm degradation during initial stages of germination, in Reference 1c, p.162.
73. **King, R. W. and Gale, M. D.**, Pre-harvest assessment of potential α-amylase production, *Cereal Res. Commun.*, 8, 157, 1980.
74. **Mares, D. J.**, Rate and location of production of alpha-amylase in relation to pre-harvest sprouting in wheat, in Reference 1d, p.494.
75. **Singh, S. P. and Paleg, L. G.**, Low temperature induction of hormone insensitivity in genotypically gibberellic acid insensitive aleurone tissue, *Plant Physiol.*, 74, 437, 1984.
76. **Cornford, C., Black, M., and Chapman, J.**, Sensitivity of developing wheat grains to gibberellins and production of alpha-amylase during grain development and maturation, in Reference 1d, p.283.
77. **Marchylo, B. A. and Kruger, J. E.**, Degradation of starch granules in maturing wheat and its relationship to alpha-amylase production by the embryo, in Reference 1d, p.483.
78. **Gibbons, G. C.**, The action of plant hormones and endosperm breakdown and embryo growth during germination of barley, in Reference 1c, p.169.
79. **Kruger, J. E. and Tipples, K. H.**, Relationships between falling number, amylography viscosity and α-amylase levels in Canadian wheat, *Cereal Res. Commun.*, 8, 97, 1980.
80. **Hylinka, I.**, Amylograph mobility and liquifaction number, *Cereal Sci. Today*, 13, 245, 1968.
81. **Okamoto, K. and Akazawa, T.**, Enzymic mechanisms of starch breakdown in germinating rice seeds. VII. Amylase formation in the epithelium, *Plant Physiol.*, 63, 336, 1979.
82. **Gibbons, G. C.**, Immunohistochemical determination of the transport pathways of α-amylase in germinating barley seeds, *Cereal Res. Commun.*, 8, 87, 1980.
83. **Akazawa, T. and Hara-Nishimura, I.**, Topographic aspects of biosynthesis, extracellular secretion, and intracellular storage of proteins in plant cells, *Annu. Rev. Plant Physiol.*, 36, 441, 1985.
84. **Kruger, J. D. and Tipples, K. H.**, Some experiences with monitoring alpha-amylase levels in Canadian wheat, in Reference 1c, p.125.
85. **Moss, H. J.**, The pasting properties of some wheat starches free of sprout damage, *Cereal Res. Commun.*, 8, 297, 1980.
86. **Moss, H. J. and Stiles, C. P.**, Cultivar effects on falling number, in Reference 1d, p.534.
87. **Ringlund, K.**, Starch quality in wheat and barley at different maturity stages in relation to seed dormancy, *Cereal Res. Commun.*, 8, 193, 1980.
88. **Ringlund, K.**, Differences in falling number at constant alpha-amylase activity, in Reference 1c, p.111.
89. **Bushuk, W. and Luckow, O. M.**, Effects of sprouting on wheat proteins and baking properties, in Reference 1d, p.188.
90. **Dengate, H. N. and Meredith, P.**, Do cultivar, season and location of growth affect starch pasting?, *Staerke*, 36, 419, 1984.
91. **Dengate, H. N.**, Swelling, pasting and gelling of wheat starch, *Adv. Cereal Sci. Technol.*, 6, 49, 1984.
92. **Meredith, P.**, The logic of testing for sprout damage, in Reference 1d, p.562.
93. **McMaster, G.**, Pre-harvest sprouting in wheat — the Australian experience, in Reference 1d, p.3.
94. **Hagberg, S.**, A rapid method for determining alpha-amylase activity, *Cereal Chem.*, 37, 218, 1960.
95. **Hagberg, S.**, Note on a simplified rapid method for determining alpha-amylase activity, *Cereal Chem.*, 38, 202, 1961.
96. **Perten, H.**, Application of the falling number method for evaluating alpha-amylase activity, *Cereal Chem.*, 41, 127, 1964.
97. **Tipples, K. H.**, A note on sample size error in the falling number method, *Cereal Chem.*, 48, 85, 1971.
98. **Mailhot, W. C.**, The falling number method, *Cereal Foods World*, 24, 50, 1980.
99. **Perten, H.**, Factors influencing falling number values, *Cereal Sci. Today*, 12, 1967.
100. **Noll, J. S.**, Effect of phytate, pH and acid treatment on falling number of sound and weathered wheats, *Cereal Chem.*, 62, 22, 1985.
101. **Lorenz, K. and Wolt, M.**, Effect of altitude on falling number values of flours, *Cereal Chem.*, 58, 80, 1981.
102. International Association for Cereal Chemistry, Method 107, Method 108, Standard Methods of the IACC, Verlag Moritz Schäfer, Detmold, F.R.G., 1968.

103. International Organization for Standardization, International standard ISO 3093, The secretariat, Magyar Szabvanzugzi Hivatal, Budapest, 1974.
104. American Association of Cereal Chemists, Method 56-81B, Method 22-07, in *Approved Methods of the AACC*, 7th ed., American Association of Cereal Chemists, St. Paul, Minn., U.S., 1972.
105. **Meredith, P.**, Inactivation of cereal α-amylase by acidification. The pasting strength of wheat flour, *Cereal Chem.*, 47, 492, 1970.
106. **Finney, P. L.**, Effect of wheat variety on the relationship between falling numbers and alpha-amylase activity, *Cereal Chem.*, 62, 258, 1985.
107. **Tara, K. A. and Bains, G. S.**, A note on the relation of sample size to Hagberg falling number values, *Cereal Chem.*, 53, 28, 1976.
108. **Allen, H. M. and Glennie-Holmes, M.**, Wheat flour with anomalous falling number/amylograph relationship, *Proc. 33rd Annu. Conf., RACI*, Cereal Chemistry Division, Royal Australian Chemistry Institute, Brisbane, p.58, 1983.
109. **Perten, H.**, A modified falling number method suitable for measuring both cereal and fungal alpha-amylase activity, *Cereal Chem.*, 61, 108, 1984.
110. **Perten, H.**, Shortening falling number analysis time for measuring the sprout damage of wheat at harvest, *Cereal Chem.*, 62, 474, 1985.
111. **Ross, A., Orth, R. A.,a nd Wrigley, C. W.**, Rapid screening for weather-damaged wheat, in Reference 1d, p.577.
112. **Meredith, P.**, The role of the amylograph in assessing sprout damage, in Reference 1c, p.119.
113. **Shuey, W. C. and Tipples, K. H.**, *The Amylograph Handbook, American Association of Cereal Chemists*, St. Paul, Minn., 1980.
114. **Voisey, P. W., Paton, D., and Timbers, G. E.**, The Ottawa starch viscometer. A new instrument for reserach and quality control applications, *Cereal Chem.*, 54, 534, 1977.
115. **Dengate, H. N. and Meredith, P.**, Wheat starch pasting measured with a "minipaster", *Staerke*, 36, 200, 1984.
116. **Guilbault, G. G. and Reitz, E. G.**, Enzymatic fluorometric assay of α-amylase in serum, *Cin. Chem.*, 22, 1702, 1976.
117. **Marchylo, B. A. and Kruger, J. E.**, A sensitive automatic method for determination of α-amylase in wheat flour, *Cereal Chem.*, 55, 188, 1978.
118. **Kruger, J. E.**, Nephelometric determination of alpha-amylase, in Reference 1d, p.584.
119. **Campbell, J. A.**, A new method for detection of sprout damaged wheat using a nephelometric determination of alpha-amylase activity, *Cereal Res. Commun.*, 8, 107, 1980.
120. **Kruger, J. E. and Tipples, K. H.**, Modified procedure for use of the Perkin-Elmer 191 Grain Amylase Analyzer in determining low levels of α-amylase in wheats and flours, *Cereal Chem.*, 58, 271, 1981.
121. **Rinderknecht, H., Wilding, P., and Haverbach, B. J.**, A new method for the determination of α-amylase, *Experientia*, 23, 805, 1967.
122. **Meredith, P. and Pomeranz, Y.**, Sprouted grain, in *Advances in Cereal Science and Technology*, Vol. 7, Pomeranz, Y., Ed., American Association of Cereal Chemists, St. Paul, 1985, 239.
123. **Barnes, W. C. and Blakeney, A. B.**, Determination of cereal alpha-amylase using a commercially available dye-labelled substrate, *Staerke*, 26, 193, 1974.
124. **Mathewson, P. R., Fahrenholz, C. H., Booth, G. D., Pomeranz, Y., and Miller, B. S.**, Results of collaborative testing using a simplified, rapid colorimetric alpha-amylase assay for evaluation of sprouted wheat, *Cereal Chem.*, 59, 108, 1982.
125. **Hejgaard, J. and Gibbons, G. C.**, Screening for α-amylase in cereals. Improved gel diffusion assay using a dye-labelled starch substrate, *Carlsberg Res. Commun.*, 44, 21, 1979.
126. **Gibbons, G. C.**, On the sequential determination of α-amylase transport and cell wall breakdown in germinating seeds of *Hordeum vulgare*, *Carlsberg Res. Commun.*, 45, 177, 1980.
127. **Jensen, Sv. A. and Heltved, F.**, An improved method for the determination of pre-germinated grains in barley, *Carlsberg Res. Commun.*, 48, 1, 1983.
128. **Jensen, Sv. A. and Law, D. P.**, A rapid method for the detection of sprouting in populations of wheat kernels, *Cereal Chem.*, 60, 406, 1983.
129. **Henry, R. J. and McLean, B. J.**, Re-evaluation of fluorescein dibutyrate staining as a method for assessment of pre-harvest sprouting in wheat and barley, *J. Cereal Sci.*, 4, 51, 1986.
130. **Jensen, Sv. A., Munck, L., and Kruger, J. E.**, A rapid fluorescence method for assessment of pre-harvest sprouting of cereal grains, *J. Cereal Sci.*, 2, 187, 1984.
131. **Corr, J. G. and Spillane, P. A.**, Some factors affecting amylolytic behaviour in Irish flours, *J. Sci. Food and Agric.*, 20, 638, 1969.
132. **Belderok, B.**, Seed dormancy problems in cereals, *Field Crop Abstr.*, 21, 203, 1968.
133. **Grahl, A. and Schrödter, H.**, H. Auswuchsvorkersage bei Weisen, *Z. Acker. u Pflanzenbau*, 137, 233, 1973.

134. **Olsson, G. and Mattson, B.,** Seed dormancy in wheat under different weather conditions, *Cereal Res. Commun.*, 4, 181, 1976.
135. **Lullukka, U.,** The effect of the temperature during the period prior to ripening on sprouting in the ear in rye and wheat varieties grown in Finland, *Cereal Res. Commun.*, 4, 93, 1976.
136. **Reiner, L. and Loch, V.,** Forecasting dormancy in barley - Ten years experience, *Cereal Res. Commun.*, 4, 107, 1976.
137. **Nielson, M. T.,** The Influence of Climatic Factors, Floral Parts, and Gibberellic Acid on Sprouting in Wheat (*Triticum aestivum* L.), Ph.D. thesis, Kansas State University, Manhattan, 1982.

INDEX

A

B

C

N

O

P

Q

R

Y